The Flock Book of Shropshire Sheep
Volume 1

by Shropshire Sheep Breeders Association

with an introduction by Jackson Chambers

This work contains material that was originally published in 1883.

This publication is within the Public Domain.

This edition is reprinted for educational purposes
and in accordance with all applicable Federal Laws.

Self Reliance Books

Get more historic titles on animal and stock breeding, gardening and old fashioned skills by visiting us at:

http://selfreliancebooks.blogspot.com/

Introduction

I am pleased to present yet another practical title on breeding and raising livestock.

The work is in the Public Domain and is re-printed here in accordance with Federal Laws.

As with all reprinted books of this age that are intended to perfectly reproduce the original edition, considerable pains and effort had to be undertaken to correct fading and sometimes outright damage to existing proofs of this title. At times, this task is quite monumental, requiring an almost total "rebuilding" of some pages from digital proofs of multiple copies. Despite this, imperfections still sometimes exist in the final proof and may detract from the visual appearance of the text.

I hope you enjoy reading this book as much as I enjoyed making it available to readers again.

Jackson Chambers

Shropshire Sheep Breeders' Association and Flock Book Society.

ADDITIONAL ERRATA.

LIFE MEMBER.—T. F. Cheatle, Dosthill, Tamworth.

PAGE 78.—Ram No. 1168, for s by Young Sultan (117), *read* s by Eaton Sultan (Pulley)

ERRATA.—VOLUME I.

——o——

NOTE.—In publishing this List of Errata the Editing Committee desire to impress upon Breeders the great importance of their writing Names and Figures *very plainly*, and of carefully checking the accuracy of their entries before sending them for insertion in the "Flock Book."

The Committee will be glad if those Breeders who may notice inaccuracies will immediately forward particulars thereof to the Secretaries, with a view to their correction.

Page 8 for De Broke *read* Lord Willoughby de Broke.

,, 36 Ram No. 314, for Bryd *read* Byrd.
,, 36 ,, 315, for Bryd *read* Byrd.
,, 43 ,, 459, for Messrs. Coxon *read* J. Coxon.
,, 47 ,, 547, for l. in 1881 *read* l. in 1880.
,, 49 ,, 596, General is the same Ram as The General (1186).
,, 52 ,, 637, for l. in 1881 *read* l. in 1880.
,, 54 ,, 685, for s Calcot (318) *read* s Calcot 2nd (318).
,, 54 ,, 693, for d by Lord Clifden (44) *read* Son of Lord Clifden (1141).
,, 57 ,, 743, for l. in 1879 *read* l. in 1880.
,, 60 ,, 804, for d by Little Lord (718) *read* d by Landseer (700).
,, 62 ,, 838, for Coxcomb (Bradburne) *read* Coxcomb (457).
,, 67 ,, 958, for d by Challenge (347) *read* d by Challenger (348).
,, 71 ,, 1018, for l. in 1867 *read* l. in 1869.
,, 75 ,, 1105, for Sir Probert *read* Sir Robert.
,, 75 ,, 1108, for l. in 1877 *read* l. in 1876.
,, 79 ,, 1191, for d by Mansell's No. 36, 1859 (797) *read* d by Mansell's No. 6, 1859 (797).
,, 80 ,, 1215, for Thomas' No. 17, 1876 *read* Thomas' No. 17 (1880).
,, 81 ,, 1222, for l. in 1879, s Grand Chief (618), *read* l. in 1880, s Standard (1159).
,, 84 ,, 1290, for l. in 1872 *read* l. in 1874.
,, 96 ,, for Shepstone *read* Swepstone.

RAM OMITTED.

LORD SALISBURY, (1192), l. in 1859 ; br. E. Holland, s Earl of Salisbury (2), d a Holland Ewe.

SHROPSHIRE SHEEP.

In establishing a Flock Book for SHROPSHIRE SHEEP the Breeders require no lengthened statement of the history, attributes, or merits of the breed in general, or of individual flocks in particular. Their object is to secure in the future absolute purity of lineage to any flocks or animals that are designated Shropshires. Much has been written respecting the Shropshires, and their celebrity for Wool and Mutton are adverted to by various writers of times long passed. Smith, in his History of Wool and Woollen Manufactures (Chron. Rusticum, published 1641), quotes the prices of English Wools in 1341, as follows :—

	To the Staple for Home Use.				For Exportation.	
	Per Sack.	Per Stone.	Per Sack.	Per Stone.	Per Sack.	Per Stone.
	£ s. d.	s. d.	£ s. d.	s. d.	£ s. d.	s. d.
Salop	6 6 4	5 0	7 6 4	5 9	9 6 4	7 3¼
Do. Staffs. including Leicester	5 6 8	4 2	6 6 8	4 11	8 6 8	6 5¼
Nottingham	4 13 4	3 7	5 13 4	4 4	7 13 4	5 10½
York and Rutland ...	4 10 0	3 5	5 10 0	4 2	7 10 0	5 9
Derby	3 3 4	2 5	4 3 3	3 2	6 3 4	4 8¼
Cumberland and West- moreland	2 13 4	2 1	3 13 4	2 10	5 13 4	2 4½

Again, Anderson in his origin of Commerce, gives the following comparative prices of English Wools for exportation in 1343—Shropshire, £9 6s. 8d. per Sack, Oxford and Staffordshire, £8 13s. 4d., Leicester, Hereford, and Gloucester, £8, Cornwall, £4., and other writers give many interesting particulars which it is unnecessary to recite in this preface.

Plymley, writing on the Agriculture of Shropshire in 1803, ascribes the Longmynds as the habitat of the ancient Shropshire Sheep, while Professor Wilson, in his Essay on the various breeds of

Sheep in Great Britain, speaks of Morfe Common, near Bridgnorth, as being a large tract of Land occupied by this hardy and indigenous race of Sheep.

The originality of the breed as one of great value is therefore abundantly proved, and there is no reason to doubt that it was spread generally over the hilly ranges and uplands of Shropshire, and, though not absolutely identical, a very similar, well-known and equally valuable race, upon which many of the Staffordshire flocks have been established, ranged the unenclosed pastures of Cannock Chase in that county. Generally speaking, no attempt was made to keep accurate Flock Books until the first recognition of the breed in the Show Yard of the Royal Agricultural Society of England at Gloucester in 1853, which was due to the instrumentality of MR. W. G. PREECE, of Shrewsbury, and the HON. ROBERT HENRY CLIVE, M.P., the latter of whom there offered Special Prizes for Shropshires. Still, the reputation of flocks in different parts of the county for particular characteristics and attributes of excellence, was well known to the Agriculturists of that and earlier periods as I have often heard confirmed from the lips of contemporary breeders of the past generation. It has been stated that certain crosses of South Down blood were introduced by some breeders prior to that period; be that as it may, no one who has achieved any success, reputation, or acknowledgment as a Shropshire breeder, has deviated from a line of pure breeding since that time. Founded on natural characteristics, it is to the good judgment in selection on the part of the majority of the breeders, that the Shropshires have obtained their present well-known notoriety for hardihood, fecundity, excellence of quality both of Wool and Mutton, and early maturity. At the same time they carry so large a proportion of lean meat to fat that Shropshire Sires are now largely used for crossing purposes in all parts of the world.

J. BOWEN-JONES.

Ensdon House, Shrewsbury,
    March 1st, 1883.

LIST OF BREEDERS,

Adney, G., Harley
Allen, G., Yew Tree Farm
Allen, G., Knightley Hall
Allsopp, Sir Henry, Hindlip Hall
Andrews, W., Nobold
Aylesford, Earl of, Packington Hall

Bach, F., Onibury
Baker, Mrs. A., Grendon
Baker, W., Moor Barns
Barber, R., Harlescote
Barrs, J. A., Nailstone
Barrs, Mrs. M., Odstone Hall
Beach, Joseph, The Hattons
Beach, Mrs. S., The Hattons
Bostock, E., Dunston Farm
Bowen-Jones, J., Ensdon House
Bowen Pryce, W., Shrawardine
Bradburne, J. H., Pipe Place
Bromley, Lea Hall
Bromley, R., Felton Butler
Byrd, C., Littywood
Byrd, S., Leese Farm

Cheatle, T. F., Dosthill
Chesham, Lord, Latimer
Clare, W. H., Twycross
Claridge, W. P., Pitchford
Clarke, C. F., Perton Grove
Coleman, E. M., Oakenshaw
Cooke, G., Horse Heath Park
Coxon, J., Freeford
Crane, E., Shrawardine
Crane, J. and E., Shrawardine
Crane, Joseph, Calcott
Crane and Tanner, Shrawardine

Darling, John, Beaudesert
Dartmouth, Earl of, Patshull

Davies, D. R., Agden Hall
Dawes, W. M., New House
De Broke, Lord Willoughby, Compton Verney
Dicken, T., Ellerdine
Dyott, Col. R., Freeford

Edwards, R., Udlington
Evans, John, Uffington
Everall, Peter, Uckington

Falmouth, Viscount, Tregothnan
Farmer, J. E., Felton
Fenn and Harding, Stone Brook House, and Wootton
Fenn, T., Stonebrook House
Firmstone, W. F., Hagley
Foster, W. O., Apley Park

German, G., Normanton
German, W., Measham Lodge
Gibson, F., Woolmet
Graham, G., Yardley
Green, J. B. and G. H., Marlow Lodge
Gretton, F., Bladon House
Griffin, H., Pell Wall
Griffiths, G., Old Hall
Groves, R. V. C., Berrington

Hall, Jos., Birchdale House
Hamar, J., Hopton Castle
Hamilton, C. W., Hamwood
Hand, Jas., Wigley
Harding, J., Wootton
Harward, John, Winterfold
Hawkins, E., Dinthill
Haydock, Captain J. B., Wootton Hall
Hill, Rev., Noel, Berrington
Hill, Hon. T. Noel, Cronkhill
Holland, E., Dumbleton Hall
Horley, T., The Fosse
Horton, Thos., Harnage Grange

Instone, E., Bourton Grange

Jefferson, R., Preston Hows
Jefferson, S., Preston Hows
Johnson, E., M.P., Farringdon House
Johnson, The Hermitage
Juckes, Tern

Keeling, C. R., Yew Tree Farm
Kenyon, The Hon. E., Maesfen
Knowles, R. M, Colston Bassett

Lander, W., Lee Hall
Lee, Henry, Ensdon
Lee, Henry and Son, Ensdon
Lowe, H., Comberford
Loder, R., M.P., Whittlebury
Lythall, E., Radford Hall

Mansell, T. J., Dudmaston Lodge
Mansell, T., Harrington Hall
Matthews, Henry, Montford
May, G. A., Elford Park
Meire, T. L., Eyton-on-Severn
Meire, S., Berrington
Meredith, E., Rednal
Meredith, R., Rednal
Minor, A. H., Astley House
Minton, J. W., Forton
Minton, T. S., Montford

Naper, J. L., Loughcrew
Nash, R. J., Park House
Nightingale, V. E., Burway
Nock, T., Sutton Maddock

Oliver, R. E., Sholebroke Lodge

Pickering, John, Alston
Pickin, W., Hilton
Pilgrim, S. C., The Outwoods
Pitt, W., Posenhall
Portland, Duke of, Clipstone Park
Price, G. C., Acton Hill
Pulley, J., M.P., Lower Eaton

Randell, C. Chadbury
Rogers, B., Ley Weobley
Rogers, H., Cheswell Grange
Rose, Miss, Mullaghmore
Ryland Thos., Pheasey Farm

Sheldon, H. J., Brailes House
Shrewsbury, Earl of, The Birches
Smith, Mrs. Henry, Sutton Maddock
Smith, Henry, Sutton Maddock

Smythe, Sir C. F. Bart., Acton Burnell
Strathmore, Earl of, Glamis Castle
Stubbs, John, Burstone

Thomas, R., The Buildings
Thornton, E., Pitchford
Timmis, Chas., Gainsboro' Hill
Townshend, Captain, Caldicote Hall
Turner, A. P., Strangworth

Vaughan, W. B., The Burway
Vivian, Sir H., Bart., Parkle Breos

Wadlow, C., Haughton
Wadlow, C., Stone Acton
Wadlow, Mrs., Haughton
Walker, B., Odstone Hill
Ward, W., Shrawardine Castle
Wenlock, Lord, Escrick Park
Wheeler, G. W., Posenhall
Williams, M., Hales
Williams, M., Jun., Bishton Hall
Wynn, Sir W. W., Bart., Wynnstay

Yates, W., Grindle House

Zetland, The Earl of, Aske

§HROPSHIRE ℝAMS

*Which have obtained Prizes or the position of Reserve Number at
the past Meetings of the Royal Agricultural Society of England.*

ABBREVIATIONS.—Breeder, Br. Exhibitor, Ex.

N.B.—Name in parentheses denotes the Breeder of the Ram immediately
preceding it.

GLOUCESTER, 1853.

AGED RAM CLASS (open short-woolled).

Prize, £10.—Shropshire Ram, (1), lambed in 1851. Ex. & Br.
Mr. Thomas Horton, Harnage Grange, near Shrewsbury.

SALISBURY, 1857.

SHEARLING RAM CLASS (short-woolled).

1st Prize, £25.—Earl of Salisbury (2), sire, Buckskin (Adney).
Ex. & Br. Mr. George Adney, Harley, Much Wenlock, Salop.

AGED RAM CLASS (short-woolled).

1st Prize, £25.—Magnum Bonum (3). Ex. Mr. Samuel Meire,
Castle Hill, Much Wenlock, Salop.

2nd Prize, £15.--Patentee (4), lambed in 1854, sire, Buckskin
(Adney). Ex. & Br. Mr. George Adney, Harley, Much
Wenlock, Salop.

CHESTER, 1858.

SHEARLING RAM CLASS.

1st Prize, £25.—Shearling Ram (5), sire, " Perfection ;" dam, by
Magnum Bonum (3). Ex. & Br. Mr. William Orme Foster,
Kinver Hill, Stourbridge.

2nd Prize, £10.—Celebrity (6), sire, Tern (1176); dam, a Crane
Ewe. Ex. & Br. Messrs. J. & E. Crane, Shrawardine,
Montford Bridge, R.S.O., Salop.

AGED RAM CLASS (open short woolled).

1st Prize £20.—Chester Billy (7), lambed in 1856, sire, Old B.
(920); dam, an Adney Ewe. Ex. & Br. Mrs. Anne Baker,
Grendon, Atherstone, Warwickshire.

2nd Prize, £10.—" Patentee " (4). Ex. Mr. George Adney.

AGED RAM CLASS.

1st Prize, £20.—" Chester Billy " (7). Ex. Mrs. Anne Baker.

2nd Prize, £10.—" Earl of Salisbury " (2). Ex. Mr. George
Adney.

WARWICK, 1859.
SHEARLING RAM CLASS.

1st Prize, £20.—" Juvenile" (8), sire, Veteran (1255); dam, by
Son of Magnum Bonum. Ex. & Br. Mr. John Coxon,
Freeford, Lichfield.

2nd Prize, £10.—" Earl of Warwick " (9), sire, ram purchased
from Mr. George Adney, in 1856. Ex. & Br. Mr. Henry J.
Sheldon, Brailes House, Shipston-on-Stour, Warwickshire.

3rd Prize, £5.—Shearling Ram (10). Ex. & Br. Mr. Thomas
Horley, The Fosse, Leamington.

AGED RAM CLASS (open short woolled).

3rd Prize.—Shropshire Ram (11), lambed in 1857. Ex. & Br. Mr.
George Adney, Harley, Much Wenlock, Salop.

AGED RAM CLASS.

1st Prize £20.—" Grendon" (12) lambed in 1857, sire, a Ram of
Mr. S. Byrd's Ex. & Br. Mrs. Anne Baker, Grendon, Atherstone, Warwickshire.

2nd Prize £10.—Celebrity (6). Ex. Messrs. J. & E. Crane.

3rd Prize £5.—Earl of Salisbury (2). Ex. Mr. George Adney.

CANTERBURY, 1860.
SHEARLING RAM CLASS.

1st Prize £15.- " Duke of Kent " (13). Ex. & Br. Mr. Thomas
Horton, Harnage Grange, near Shrewsbury.

2nd Prize £5.—Commonwealth (14), sire, Teru (1176); dam, a
Crane Ewe. Ex. & Br. Messrs. J. & E. Crane, Shrawardine,
Montford Bridge, R.S.O., Salop.

AGED RAM CLASS.

1st Prize £15.—"Canterbury Patentee" (15), lambed in 1858, sire, Patentee (4). Ex. & Br. Mr. Sampson Byrd, Leese Farm, Stafford.

2nd Prize £5.—Shropshire Ram (16). Ex. & Br. Mr. Thomas Horton, Harnage Grange, near Shrewsbury.

LEEDS, 1861.
SHEARLING RAM CLASS.

1st Prize £15.—Lord Suffolk (17). Ex. & Br. Mr. Thos. Horton, Harnage Grange, near Shrewsbury.

2nd Prize £10.—Shearling Ram (18). Ex. & Br. Messrs. J. & E. Crane, Shrawardine, Montford Bridge, R.S.O., Salop.

3rd Prize £5.—Shearling Ram (19). Ex. & Br. Mrs. Anne Baker, Grendon, Atherstone, Warwickshire.

AGED RAM CLASS.

1st Prize, £15.—Patentee 2nd (20), lambed in 1858, sire, Patentee (4). Ex. Mr. Edward Holland, Dumbleton Hall, Evesham. Br. Mr. S. Byrd, The Leese Farm, Stafford.

2nd Prize, £10.—Shropshire Ram (21), lambed in 1859. Ex. & Br. Mr. Thos. Horton, Harnage Grange, near Shrewsbury.

3rd Prize, £5.— Shropshire Ram (22), lambed in 1858. Ex. & Br. Mr. Thos. Horley, The Fosse, Leamington.

BATTERSEA, 1862.
SHEARLING RAM CLASS.

1st Prize, £15.—"Lord Salop" (23) sire, Duke of Kent (13). Ex. & Br. Mr. Thomas Horton, Harnage Grange, near Shrewsbury.

2nd Prize, £10.—"John Bull" (24), sire, Young Buckskin, (1291). Ex. & Br. Mr. Thomas Mansell, Adcott Hall, Shrewsbury.

3rd Prize, £5.—Shearling Ram (25). Ex. & Br. Mr. Henry Matthews, Montford, Montford Bridge, R.S.O., Salop.

Reserve Number.—Black Prince 2nd (26), sire, Black Prince. dam by Masfen's No. 6, 1854. Ex. & Br. Mr. Thomas Horley, The Fosse, Leamington.

AGED RAM CLASS.

1st Prize, £15.—"Havelock" (27), lambed in 1859, sire, Lord Harley (Adney); dam by Masfen's No. 6, 1854. Ex. & Br. Mr. Thos. Horley, The Fosse, Leamington.

2nd Prize, £10.—Patentee the 4th, alias Patentee the Prime (28), lambed in 1858, sire, Patentee (4).   Ex. Mr. Pryce W. Bowen, Shrawardine, Montford Bridge R.S.O.   Br. Mr. S. Byrd, The Leese Farm, Stafford.

3rd Prize, £5.—"St. Patrick" (29) lambed in 1859, sire, Old Shropshire.   Ex. & Br. Mr. Thomas Horton, Harnage Grange, near Shrewsbury.

Reserve Number.—"Tommy No. 1" (30), lambed in 1860, sire, "Old Sheep," dam, "Queen Bet 2nd,".   Ex. & Br. The Right Hon. Lord Wenlock, Bourton Cottage, Much Wenlock, Salop.

WORCESTER, 1863.
SHEARLING RAM CLASS.

1st Prize, £20.—"Earl of Shrewsbury" (31).   Ex. & Br. Mr. John Stubbs, Weston Hall, Stafford.

2nd Prize, £10 —"Quality," (32) sire, "Duke of Kent" (13).   Ex. & Br. Mr. Thos. Horton, Harnage Grange, near Shrewsbury.

3rd Prize (Silver Medal).—"Capt. Semmes (33) sire, "Sir Henry." dam by Lord Suffolk (17).   Ex. and Br. Mr. Thomas Horton, Harnage Grange, near Shrewsbury.

Reserve Number.— Shearling Ram (34).   Ex. and Br. Mr. W. O. Foster, Kinver Hill, near Stourbridge.

AGED RAM CLASS.

1st Prize, £20.—"Worcester Patron," (35), lambed in 1861, sire Preece's Patron ; dam by Comet (Crane), gr. d. by Chester Billy (7).   Ex. & Br. Mr. Pryce W. Bowen, Shrawardine Castle, Montford Bridge, R.S.O.

2nd Prize, £10.—"Pattern" (36), lambed in 1861, sire, Patent (931) ; dam, a Dyott Ewe.   Ex. & Br. Col. R. Dyott, Freeford Hall, Lichfield.

3rd Prize, Silver Medal.—"Nobleman," (37), lambed in 1860, sire, Patent (931) ; dam, by Old Packington (Coxon).   Ex. & Br. Mr. John Coxon, Freeford, Lichfield.

Reserve Number.—"Magic" (38), lambed in 1860, sire Old Cross dam by Goldfinder (Adney).   Ex. & Br. Mr. R. H. Masfen, Pendeford, Wolverhampton.

NEWCASTLE-ON-TYNE, 1864.
SHEARLING RAM CLASS.

1st Prize, £20.—"Duke of Newcastle" (39,) sire, Pride of Pitchford (961) ; dam, by Canterbury Patentee 2nd.(Thornton) Ex. & Br. Mr. Edward Thornton, Pitchford, Shrewsbury.

2nd Prize, £10.—Shearling Ram (40). Ex. & Br. Mr. Henry Matthews, Montford, Montford Bridge, R.S.O., Salop.

3rd Prize, Silver Medal.—" Novelty" (41), sire, Nobleman (37); dam by Son of Magnum Bonum. Ex. & Br. Mr. John Coxon, Freeford, Lichfield.

Reserve Number.—" Patentee 5th" (42), sire, Constitution; dam by Patentee (4). Ex. & Br. Mr. Sampson Byrd, Leese Farm, Stafford.

AGED RAM CLASS.

1st Prize, £20.—" Black Knight" (43), lambed in 1862, sire, Valiant the 2nd; dam by Lord Flash (747). Ex. & Br. Mr. John Coxon, Freeford, Lichfield.

2nd Prize, £10.—" Lord Clifden" (44), lambed in 1862, sire, Comet (Crane); dam by Patentee the Prime (28). Ex. & Br. Mr. Pryce W. Bowen, Shrawardine Castle, near Shrewsbury.

Reserve Number.—Shropshire Ram (45), lambed in 1860. Ex. Mr. Joseph Simpson, Spofforth Park, Wetherby, Yorks. Br. Lord Wenlock.

PLYMOUTH, 1865.

SHEARLING RAM CLASS.

1st Prize, £20.—"Mansion" (46), sire, Maccaroni (785); dam by Young Buckskin (1291). Ex. & Br. Mr. Thomas Mansell, Adcott, near Shrewsbury.

2nd Prize, £10.—"Plymouth Prize" (47), sire, Nobleman (37) dam by Tern (1176). Ex. & Br. Messrs. J. & E. Crane, Shrawardine, Montford Bridge, R.S.O., Salop.

3rd Prize, £5.—" Earl of Plymouth" (48). sire, Maccaroni (785); dam by Laurel (T. Mansell). Ex. & Br. Mr. Thomas Mansell, Adcott, near Shrewsbury.

Reserve Number.—" Big Plymouth" (49), sire, Chieftain (384); dam by Celebrity (6). Ex. & Br. Messrs. J. & E. Crane, Shrawardine, Montford Bridge, R.S.O.

AGED RAM CLASS.

1st Prize, £20.—" Beaufort" (50), lambed in 1863, sire, Pattern (36) dam, a Dyott Ewe. Ex. & Br. Col. R. Dyott, Freeford, Lichfield.

2nd Prize, £10.—" Volunteer" (51) lambed in 1863, sire, Pride of Pitchford (961), dam, by Canterbury Patentee 2nd (Thornton). Ex. & Br. Mr. Edward Thornton, Pitchford, Shrewsbury.

3rd Prize, £5.—"Young Quality" (52), lambed in 1861, sire, Quality; dam by Patentee (4). Ex. & Br. Mr. Sampson Byrd, Leese Farm, Stafford.

Reserve Number.—"Lord Clifden" (44). Ex. Mr. Pryce W. Bowen.

No Show in 1866, (Cattle Plague year)

BURY-ST.-EDMUNDS, 1867.

SHEARLING RAM CLASS.

1st Prize, £20.—"Cardinal" (53), sire, Nonpareil (908); dam by Young Emperor, (1305); gr d by Humphrey Davy (676); g gr d by Emperor (525). Ex. & Br. Mr. John Evans, Uffington, Shrewsbury.

2nd Prize, £10.—" Abbot of Bury " (54), sire, Nonpareil (908); dam by Emperor (525), gr d by Humphrey Davy (676); g gr d by Own Brother to Bossy (925). Ex. & Br. Mr. John Evans, Uffington, Shrewsbury.

3rd Prize, £5.—" Mansion 2nd" (55), sire, Earl of Plymouth (48), dam by Young Buckskin (1291). Ex. & Br. Mr. Thomas Mansell, Adcott, near Shrewsbury.

Reserve Number.—" Lord Uffington" (56), sire, Nonpareil (908); dam by Emperor (525); gr d by Humphrey Davy (676); g gr d by Own Brother to Bossy (925.) Ex. & Br. Mr. John Evans, Uffington, Shrewsbury.

AGED RAM CLASS.

1st Prize, £20.—' Corsair " (57), lambed in 1865, sire, Chieftain, (384), dam by Patentee (4). Ex. & Br. Messrs. J. & E. Crane, Shrawardine, Montford Bridge, R.S.O., Salop.

2nd Prize, £10.—Bonnyface (58), lambed in 1865, sire, Lord Clifden (44), dam by a Pryce W. Bowen Ram. Ex. & Br. Mr. Thomas Mansell, Adcott, near Shrewsbury.

3rd Prize, £5.—"Emigrant" (59), lambed in 1864, sire, Pride of Pitchford (961); dam by Emperor (525); gr d by Humphrey Davy (676); g gr d by Own Brother to Bossy (925). Ex. & Br. Mr. John Evans, Uffington, Shrewsbury.

Reserve number.—"Great Eastern" (60), lambed in 1864, sire, Picture (S. Bryd), dam by Veteran (1255). Ex. & Br. Mr. Geo. Anderson May, Elford Park, Tamworth.

LEICESTER, 1868.

SHEARLING RAM CLASS.

1st Prize, £20.—" Earl of Leicester " (61), sire, Young Celebrity (Crane); dam by Tern (1176). Ex. & Br. Messrs. J. & E. Crane, Shrawardine, Montford Bridge, R.S.O., Salop.

2nd Prize, £10.—Shearling Ram (62), sire, Conservative (435); dam by Maccaroni (785). Ex. & Br. Mr. Thomas Mansell, Adcott, near Shrewsbury.

3rd Prize, £5.—Shearling Ram (63), sire, Conservative (435); dam by Maccaroni (785). Ex. & Br. Mr. Thomas Mansell, Adcott, near Shrewsbury.

Reserve Number.—" Pride of Hanmer " (64), sire, Pride of Agden (Jones) ; dam by Commonwealth (14). Ex. & Br. Mr. Geo. Griffiths, Old Hall, Hanmer, Whitchurch, Salop.

AGED RAM CLASS.

1st Prize, £20.—" Lord Uffington," (56). Ex. Mr. J. Evans.

2nd Prize, £10.—" Crosswood Hero " (65), lambed in 1866, sire, Chieftain (384); dam by Young Tern (Crane). Ex. & Br. Messrs. J. & E. Crane, Shrawardine, Montford Bridge, R.S.O., Salop.

3rd Prize, £5.—" Viceroy " (66), lambed in 1865, sire, Grindle (Masfen) ; dam by Model. Ex. Mr. Charles Byrd, Littywood, Stafford. Br. Mr. C. R. Keeling, Yew Tree Farm, Penkridge, Stafford.

Reserve Number.—" Royal Ranger " (67), lambed in 1866, sire, Ranger (Bradburne) ; dam by Worcester Royal Special. Ex. & Br. Mr. Edward Holland, Dumbleton Hall, Evesham, Worcestershire.

MANCHESTER, 1869.

SHEARLING RAM CLASS.

1st Prize, £20.—Shearling Ram, (68), sire, Old Latimer (H. Smith); dam a Crane Ewe. Ex. & Br. The Right Hon. Lord Chesham, Latimer, Chesham, Bucks.

2nd Prize, £10.—" Chancellor " (69), sire, Commander (415); dam by Cavalier. Ex. & Br. Mr. John Coxon, Freeford, Lichfield.

3rd Prize, £5.—" Duke of Manchester" (70), sire, Cardinal (53); dam by Gayton Prince (592). Ex. Mrs. Beach, The Hattons, Brewood, Staffordshire. Br. The late Mr. J. Beach.

Reserve Number.—Shearling Ram (71), sire, Old Latimer (H. Smith); dam by H. Matthews' Ram. Ex. & Br. The Right Hon. Lord Chesham, Latimer, Chesham, Bucks.

AGED RAM CLASS.

1st Prize, £20.—" Leviathan " (72), lambed in 1866, sire, Lord Clifden, (44). Ex. & Br. Mr. Henry Matthews, Montford, Montford Bridge, R.S.O., Salop.

2nd Prize, £10.—Shropshire Ram (73), lambed in 1867. Ex. & Br. Mr. Thomas Horton, Harnage Grange, Shrewsbury.

3rd Prize, £5.—"The Duke" (74), lambed in 1867, sire, Lord Warden (777); dam by Patentee the Prime (28). Ex. Mrs. Pryce W. Bowen, Shrawardine, Montford Bridge, R.S.O., Salop. Br. Mr. Pryce W. Bowen.

Reserve Number.—"Black Prince" (75), lambed in 1867, sire, Guardian (642), dam by Leese Patentee (S. Byrd). Ex. & Br. Mr. Charles Byrd, Littywood, Stafford.

OXFORD, 1870.
Shearling Ram Class.

1st Prize, £20.—"Captivator" (76), sire, Commander (415); dam by Novelty (41). Ex· & Br. Mr. John Coxon, Freeford, Lichfield.

2nd Prize, £10.—"Oxford Hero" (77), sire, Conservative (435); dam by Short-legged Patentee (1076). Ex. & Br. Mr. Thomas Mansell, Adcott, near Shrewsbury.

3rd Prize, £5.—Shearling Ram (78), sire, Young Gayton (1307); dam a Beach Ewe. Ex. & Br. Mrs. Beach, The Hattons, Brewood, Staffordshire.

Reserve Number.—"Matthew" (79), sire, Masterman (Masfen); dam by Henry Matthew's Ram. Ex. & Br. The Right Hon. Lord Chesham, Latimer, Chesham, Bucks.

Aged Ram Class.

1st Prize £20.—"Standard Bearer" (80), lambed in 1867, sire Corsair (57); dam by Mainstay (Masfen). Ex. Mr. John Evans, Uffington, Shrewsbury. Br. Mr. R. H. Masfen, Pendeford, Wolverhampton.

2nd Prize, £10.—Shropshire Ram (81), lambed in 1868, sire, Cardinal (53); dam a Beach Ewe. Ex. Mrs. Beach, The Hattons, Brewood, Staffordshire. Br. The late Mr. J. Beach.

3rd Prize, £5.—Shropshire Ram (82), lambed in 1868, sire, Cardinal (53); dam a Beach Ewe. Ex. Mrs. Beach, The Hattons, Brewood, Staffordshire. Br. The late Mr. J. Beach.

Reserve Number.—Blood Royal (83), lambed in 1868, sire, Conservative (435); dam by Mansell's No. 6, 1859 (797). Ex. & Br. Mr. Thomas Mansell, Adcott, near Shrewsbury.

WOLVERHAMPTON, 1871.
Shearling Ram Class.

1st Prize, £20.—"Hampton Hero" (84), sire, Blood Royal (83); dam by Guardian (642), gr d by Canterbury 2nd (328). Ex. & Br. Mr. Charles Byrd, Littywood, Stafford.

2nd Prize, £10.—"True Type" (85), sire, Marquis (820); dam by Conservative (435). Ex. & Br. Mr. Thomas Mansell, Adcott, near Shrewsbury.

3rd Prize, £5.—Shearling Ram (86). Ex. Mr. John Stubbs, Burstone, Stone, Staffordshire.

Reserve number.—"Proud Salopian" (87), sire, Cardinal (53), dam by Young Emperor (1305); gr d by Pride of Pitchford, (961); g gr d by Sir Samuel (1113). Ex. & Br. Mr. John Evans, Uffington, Shrewsbury.

AGED RAM CLASS.

1st Prize, £20.—"Major" (88), lambed in 1869, sire, Conservative (435); dam by Maccaroni (785). Ex. & Br. Mr. Thomas Mansell, Adcott, near Shrewsbury.

2nd Prize, £10.—"Broadgauge" (89), lambed in 1869, sire, Nonpareil (908); dam by Young Emperor (130); gr d by Sir Samuel (1113), g gr d by Humphrey Davy (676). Ex. & Br. Mr. John Evans, Uffington, Shrewsbury.

3rd Prize, £5.—"Conductor" (90), lambed in 1869, sire, Commander (415); dam by Novelty (41). Ex. & Br. Mr. John Coxon, Freeford, Lichfield.

Reserve number.—"Blood Royal" (83). Ex. Mr. Thomas Nock, Sutton Maddock, Shifnal.

CARDIFF, 1872.

SHEARLING RAM CLASS.

1st Prize, £20.—"Marquis of Bute" (91), sire, Matthew (79); dam by H. Matthew's Ram. Ex. & Br. The Right Hon. Lord Chesham, Latimer, Chesham, Bucks.

2nd Prize, £10.—Shearling Ram (92), sire, Young Latimer. Ex. & Br. Mr. Wm. O. Foster, Kinver Hill, Stourbridge, Worcestershire.

3rd Prize, £5.—"Evans No. 8, 1872" (93), sire, Hardlines (647); dam Cardinal (53), gr d by Competition (425), g gr d by Sir Samuel (1113). Ex. & Br. Mr. John Evans, Uffington, Shrewsbury.

Reserve Number.—Shearling Ram (94), sire, Duke of Manchester (70); dam by Gayton Prince (592). Ex. & Br. Mrs. Beach, The Hattons, Brewood, Staffordshire.

AGED RAM CLASS.

1st Prize, £20.—"Legatee" (95), lambed in 1870, sire, Conservative (435); dam by Mansell's No. 6, 1859 (797). Ex. & Br. Mr. Thomas Mansell, Adcott, near Shrewsbury.

2nd Prize, £10.—" Constitution " (96 , lambed in 1870, sire, Conservative (435) ; dam by Young Courtier (T. Mansell) ; g d by Short-legged Patentee (1076). Ex. & Br. Mr. Thomas Mansell, Adcott, near Shrewsbury.

3rd Prize, £5.— Shropshire Ram (97), lambed in 1870, sire, Duke of Manchester (70) ; dam a Beach Ewe. Ex. & Br. Mrs. Beach, The Hattons, Brewood, Staffordshire.

Reserve Number.—" Lismore " (98), lambed in 1870. Ex. & Br. The Right Hon. Lord Chesham, Latimer, Chesham, Bucks.

HULL, 1873.

SHEARLING RAM CLASS.

1st Prize, £20.—"Lord Kingston " (99), sire, Grandson of Emperor, dam by Old Latimer (H. Smith). Ex. & Br. The Right Hon. Lord Chesham, Latimer, Chesham, Bucks.

2nd Prize, £10.—"Lord Hull " (100), sire, Briton (296) ; dam by Old Latimer (H. Smith). Ex. & Br. Mrs. Beach, The Hattons, Brewood, Staffordshire.

3rd Prize, £5.—" Confidence" (101), sire, Pilot (Mansell) ; dam by Mansion 3rd (818). Ex. & Br. Mr. John Coxon, Freeford, Lichfield.

Reserve Number.—Shearling Ram (102), sire, Oxford Hero (77). Ex. & Br. Mr. Charles Byrd, Littywood, Stafford.

AGED RAM CLASS.

1st Prize, £20.—" Claudius " (103) lambed in 1870, sire, Caractacus (335) ; dam by Celebrity (6). Ex. & Br. Mr. Edward Crane, Shrawardine, Montford Bridge, R.S.O., Salop.

2nd Prize, £10.—"Ensdon Hero " (104), lambed in 1871, sire, Kingscraft (698) ; dam by Novelty (41), g d by Nobleman (37). Ex. & Br. Mr. Thomas Fenn, Stonebrook House, Downton, Ludlow.

3rd Prize, £5.—" Dorchester Hero " (105), lambed in 1871, sire, T. Mansell's No. 8, 1870 ; dam by Milton (Masfen). Ex. Mr. Joseph Pulley, Lower Eaton, Hereford. Br. The Right Hon. Lord Chesham, Latimer, Chesham, Bucks.

Reserve Number.—Shropshire Ram (106), lambed in 1870, sire, Duke of Manchester (70) ; dam a Beach Ewe. Ex. & Br. Mrs. Beach, The Hattons, Brewood, Staffordshire.

BEDFORD, 1874.

SHEARLING RAM CLASS.

1st Prize, £20.—" Duke of Bedford " (107), sire, grandson of Oxford Hero (627) ; dam a Sutton Maddock ewe, Ex. & Br. The Right Hon. Lord Chesham, Latimer, Chesham, Bucks.

2nd Prize, £10.—"Raby Duke" (108), sire, Longitude (729); dam by Conservative (435). Ex. & Br. Messrs. T. & T. J. Mansell, Ercall Park, Wellington, Salop.

3rd Prize, £5.—"Sir William" (109), sire, Buckskin (307); dam by Chieftain (384). Ex. & Br. Mr. Edward Crane, Shrawardine, Montford Bridge, R.S.O., Salop.

Reserve number.—Shropshire Ram (110), sire, Briton (296); dam a Beach Ewe. Ex. & Br. Mrs. Beach, The Hattons, Brewood, Staffordshire.

AGED RAM CLASS.

1st Prize, £20.—"Bedford Hero" (111), lambed in 1872, sire, Calcot (317); dam by Conservative (435). Ex. & Br. Messrs. T. & T. J. Mansell, Ercall Park, Wellington, Salop.

2nd Prize, £10.—"Caligula" (112), lambed in 1872, sire, Cato (Crane), dam by Celebrity (6). Ex. & Br. Mr. Edward Crane, Shrawardine, Montford Bridge, R.S.O., Salop.

3rd Prize, £5.—"Paddy" (113), lambed in 1872, sire, Grandson of Duke of Manchester. Ex. & Br. The Right Hon. Lord Chesham, Latimer, Chesham, Bucks.

Reserve number.—"The Ruler" (114), lambed in 1871, sire, Son of Cardinal, dam by Young Duke (1303). Ex. Mr. Wm. German, Measham Lodge, Atherstone. Br. Mr. B. Walker, Odstone Hill, Atherstone.

TAUNTON, 1875.
SHEARLING RAM CLASS.

1st Prize, £20.—"Royal Taunton" (115), sire, Lord Vincent (Chesham); dam by a Ram bred by the Hon. Kenyon. Ex. & Br. The Right Hon. Lord Chesham, Latimer, Chesham, Bucks.

2nd Prize, £10.—"Cato" (116), sire, Crown Prince, (483), dam by Chancellor, (69). Ex. & Br. Mr. W. German, Measham Lodge, Atherstone.

3rd Prize, £5.—"Young Sultan" (117), sire, Sultan, (1167), dam by Buckskin, (307). Ex. & Br. Mr. Joseph Pulley, Lower Eaton, Hereford.

Reserve Number.—"Taunton Reserve" (118), sire, Lord Kingston (99). Ex. & Br. The Right Hon. Lord Chesham, Latimer, Chesham, Bucks.

AGED RAM CLASS.

1st Prize, £20.—"Hereford" (119) lambed in 1873, sire, Dorchester Hero (105); dam by Fat Back, (551). Ex. & Br. Mr. Joseph Pulley, Lower Eaton, Hereford.

2nd Prize, £10.—" Patrician " (120), lambed in 1873, sire, Oxford Hero (77) ; dam by Matthew (79). Ex. & Br. The Right Hon. Lord Chesham, Latimer, Chesham, Bucks.

3rd Prize, £5.—" Lord Taunton " (121), lambed in 1873, sire, Claudius (103); dam by Shamrock, (1054). Ex. & Br. Mr. E. Crane, Shrawardine, Montford Bridge, R.S.O., Salop.

Reserve Number.—" Candidate " (122), lambed in 1873, sire, Crown Prince, (483), dam by Chancellor, (69)· Ex. & Br. Mr. W. German, Measham Lodge, Atherstone.

BIRMINGHAM, 1876.

SHEARLING RAM CLASS.

1st Prize, £20.—" Lord Aston " (123), sire, Bedford Hero (111); dam a Forton Ewe. Ex. & Br. Mr. J. W. Minton, Forton, near Shrewsbury.

2nd Prize, £10.—" Royal Aston " (124), sire, Grandson of Lord Kingston ; dam by Matthew (79). Ex. & Br. The Right Hon. Lord Chesham, Latimer, Chesham, Bucks.

3rd Prize, £5.—" Aston " (125), sire, Goliah, (611) dam a Sheldon Ewe. Ex. & Br. Mr. H. J. Sheldon, Shipston-on-Stour, Warwickshire.

Reserve Number.—" Birmingham Reserve " (126), sire, Ensdon Hero (104) ; dam by Lord Kenyon, gr d by Novelty (41). Ex. & Br. Mr. Thomas Fenn, Stonebrook House, Downton, Ludlow.

AGED RAM CLASS.

1st Prize, £20.—" Example " (127), lambed in 1874, sire, Corporal (443); dam a Coxon Ewe. Ex & Br. Capt. Hy. Townshend, Caldicote Farm, Nuneaton.

2nd Prize, £10,—" Columbus " (128), lambed in 1874, sire, Claudius (103); dam by Corsair (57). Ex. & Br. Mr. Edward Crane, Shrawardine, Montford Bridge, R.S.O., Salop.

3rd Prize, £5.—" Sheldon's No. 1, 1876 " (129), lambed in 1874, sire, "Tory Peer" (1229); dam a Sheldon Ewe. Ex. & Br. Mr. H. J. Sheldon, Shipston-on-Stour, Warwickshire.

Reserve Number.—" Trojan " (130), lambed in 1874, sire, Tarquin (Masfen.) Ex. & Br. Mr. William Baker, Moor Barns, Atherstone.

LIVERPOOL, 1877.

SHEARLING RAM CLASS.

1st Prize, £20. —" Lord Liverpool " (131), sire, Example (127), dam a Masfen Ewe. Ex. & Br. Capt. Henry Townshend, Caldicote Farm, Nuneaton.

2nd Prize, £10.—"3rd Marquis of Bute" (132), sire, Marquis of Bute (91). Ex. & Br. The Right Hon. Lord Chesham, Latimer, Chesham, Bucks.

3rd Prize, £5.—"Grandeur" (133), sire, Double B, (500), dam by Little Lord (718), gr d by Rifleman, (1007). Ex. & Br. Mr. Thos. Jas. Mansell, Dudmaston Lodge, Bridgnorth, Salop.

Reserve number.—Shearling Ram (134), Sire, Marquis of Bute (91). Ex. & Br. The Right Hon. Lord Chesham, Latimer, Chesham, Bucks.

AGED RAM CLASS.

1st Prize, £20.—"Lord Harlech" (135), lambed in 1875, sire, General, (596), dam by Conservative, (435) Ex. & Br. Mr. Thomas Mansell, Ercall Park, Wellington, Salop.

2nd Prize, £10.—"Talisman" (136), lambed in 1875, sire, Sample (Chesham), dam by a J. Coxon Ram. Ex. & Br. Capt. Henry Townshend, Caldicote Farm, Nuneaton.

3rd Prize, £5.—"Oswestry Champion" (137), lambed in 1874, sire, The Warder, (1210), dam by Paddy, (929). Ex. & Br. Mr. Francis Bach, Onibury, Craven Arms, Salop.

BRISTOL, 1878.

SHEARLING RAM CLASS.

1st Prize, £20.—"Royal Bristol" (138), sire, Example (127); dam, a Masfen Ewe. Ex. & Br. Captain Henry Townshend, Caldicote Farm, Nuneaton.

2nd Prize, £10.—"Georgius" (139), sire, Prince Imperial, (966) dam a Keeling Ewe. Ex. & Br. Mr. George Graham, The Oaklands, Birmingham.

3rd Prize, £5.—"Young Duke" (140), sire, May Duke, (837), dam by Foreman, (569). Ex. & Br. Mr. Thomas Jas. Mansell, Dudmaston, Lodge, Bridgnorth, Salop.

Reserve Number.—Bristol Reserve (141), sire, "Aston" (125), dam by Lord Chesham's No. 5, 1873. Ex. & Br. Mr. Thomas Nock, Sutton Maddock, Shifnal.

AGED RAM CLASS.

1st Prize, £20.—"Talisman" (136). Ex. Captain H. Townshend.

2nd Prize, £10.—"Bristol Prize" (142), lambed in 1876, sire, Claude Duval, (394), dam by Corsair (57). Ex. & Br. Messrs. Crane and Tanner, Shrawardine, Montford Bridge, R S.O., Salop.

3rd Prize, £5.—"Sheldon's No. 2, 1878" (143), lambed in 1876, sire Goliah, (611) dam, a Sheldon Ewe. Ex. & Br. Mr. H. J. Sheldon, Brailes House, Shipston-on-Stour, Warwickshire.

Reserve Number.—" Bristol Reserve" (144), lambed in 1876, sire, Claude Duval, (394) dam by Corsair, (57). Ex. & Br. Messrs. Crane & Tanner, Shrawardine, Montford Bridge, R.S.O.,Salop.

LONDON, 1879.

SHEARLING RAM CLASS.

1st Prize, £20.—" Yardley " (145); sire, Georgius (139), dam, a T. J. Mansell Ewe. Ex. & Br. Mr. George Graham, Oaklands, near Birmingham.

2nd Prize, £10.—" Lord Mayor" (146), sire, County Member, (452), dam by Calcot, (317), gr d by Pattern, (933). Ex. & Br. Mr. Thomas James Mansell, Dudmaston Lodge, Bridgnorth.

3rd Prize, £5.—" Sir Guy " (147), sire, Sir Gray, (1096) dam by Longbow, (728). Ex. & Br. Mr. J. Lenox Naper, Loughcrew, Oldcastle, Ireland.

Reserve Number. —" Sheldon's No. 1, 1879 " (148), sire, Taunton Reserve (118). Ex. & Br. Mr. Hy. J. Sheldon, Brailes House, Shipston-on-Stour, Warwickshire.

AGED RAM CLASS.

1st Prize, £20.—" Lord Kilburn " (149), lambed in 1877, sire, Oswestry Champion (137); dam by Paddy, (929) Ex. and Br. Mr. Francis Bach, Onibury, Craven Arms, Salop.

2nd Prize, £10.—" Victor " (150), lambed in 1877, sire, Double B (500), dam by Landseer, (700), gr d by Conservative, (435). Ex. & Br. Mr. Thos. Jas. Mansell, Dudmaston Lodge, Bridgnorth, Salop.

3rd Prize, £5.—" Acton " (151), lambed in 1877, sire, Lord Acton, (731), dam by Pulley's No. 1, 1872. Ex. & Br. Mr. James Edward Farmer, Felton, Ludlow, Salop.

Reserve Number.—" Bristol Reserve " (141). Ex. Mr. Thomas Nock, Sutton Maddock, Shifnal.

CARLISLE, 1880.

SHEARLING RAM CLASS.

1st Prize, £20.—" Lord Carlisle " (153), sire, Marquis of Bath, (822), dam by Son of Little Lord, (1139). Ex. & Br. Messrs. J. W. & T. S. Minton, Montford, Montford Bridge, R.S.O., Salop.

2nd Prize, £10.—" Border Chief " (154), sire, The Czar (1182), dam by Lord Beaconsfield (736). Ex. & Br. Mrs. Maria Barrs, Odstone Hall, Atherstone.

3rd Prize, £5.—"Cumberland Hero" (155), sire, His Lordship, (669), dam by Son of Conservative, (1134). Ex. & Br. Messrs. J. W. & T. S. Minton, Montford, Montford Bridge, R.S.O., Salop.

Reserve number.—"Merry Carlisle" (156), sire, Cossack, (445), dam by Longbow, (728). Ex. & Br. Mr. J. Lenox Naper, Loughcrew, Oldcastle, Ireland.

AGED RAM CLASS.

1st Prize, £20.—"Marquis of Lorne" (157), lambed in 1878, sire, Corporal (443), dam by Major, (791). Ex. & Br. Mrs. Maria Barrs, Odstone Hall, Atherstone.

2nd Prize, £10.—"Prince Victor" (158), lambed in 1878, sire, Prince (963), dam by Benthall Chieftain (246), gr d by Mansell's No. 6, 1868, (798), g grd by The Ruler, (1204). Ex. & Br. Mr. Richard Thomas, The Buildings, Baschurch, Salop

3rd Prize, £5.—"Sir Guy" (147). Ex. Mr. J. Lenox Naper.

Reserve number.—"Royal Reserve" (159), lambed in 1878, sire, Son of Ensdon Hero (Fenn), dam a Bromley Ewe. Ex. Messrs. J. W. & T. S. Minton, Montford, Montford Bridge, R.S.O., Salop. Br. Mr. Bromley.

DERBY, 1881.
SHEARLING RAM CLASS.

1st Prize, £20.—"Montford Hero" (160), sire, Marquis of Bath, (822), dam by May Duke, (837) Ex. & Br. Mr. Thos. Stephen Minton, Montford, Montford Bridge, R.S.O., Salop.

2nd Prize, £15.—"Crown Derby" (161), sire, Pride of Bishton, (954), dam by Longbow, (728). Ex. & Br. Mr. J. Lenox Naper, Loughcrew, Oldcastle, Ireland.

3rd Prize, £10.—"Shearling Ram" (162), sire, Georgius (139). Ex. & Br. Mr. Geo. Graham, Oaklands, near Birmingham.

4th Prize, £5.—"Prince Royal" (163), sire, Royal Reserve (159); dam a Minton Ewe. Ex. & Br. Mr. Thos. Stephen Minton, Montford, Montford Bridge, R.S.O., Salop.

Reserve Number.—"Sterling" (164), sire, Milton, (843), dam by Truestock, (1243), gr d by Little Lord, (718) Ex. & Br. Mr. Thomas James Mansell, Dudmaston Lodge, Bridgnorth, Salop.

AGED RAM CLASS.

1st Prize, £20. "Dudmaston Hero" (165), lambed in 1879, sire, Pride of Montford, (959) dam by Truestock, (1243), gr d by Little Lord, (718). Ex. & Br. Mr. Thos. Jas. Mansell, Dudmaston Lodge, Bridgnorth, Salop.

2nd Prize, £15.—" Royal Reserve " (159).   Ex. Mr. T. S. Minton.

3rd Prize, £10.—" Carlisle " (166), lambed in 1878, sire, Dudmaston, (504) dam by Caligula (112).   Ex. Mr. Jas. Edward Farmer, Felton, Ludlow, Salop. Br. Messrs. Crane and Tanner, Shrawardine, Montford Bridge.

4th Prize, £5.—" Lord Oxon " (167), lambed in 1878, sire, Columbus (128); dam by Claudius (103).   Ex & Br. Messrs. Crane and Tanner, Shrawardine, Montford Bridge, R.S.O., Salop.

Reserve Number.—" Derby Reserve " (168), lambed in 1878, sire, The General (Masfen); dam by a Ram bred by the late Mr. H. Pickstock. Ex. & Br. His Grace The Duke of Portland, Clipstone Park, Mansfield, Notts.

READING, 1882.

SHEARLING RAM CLASS.

1st Prize, £15.—" Prince Regent" (169), sire, Milton, (843), dam by Beach's No. 17, 1876, (228), gr d by Calcot, (317). Ex. & Br. Mr. Thomas James Mansell, Dudmaston Lodge, Bridgnorth, Salop.

2nd Prize, £10.—" Second Best" (170), sire, Harlescote (651), dam by Truelight, (1242), gr d by Co-Monument, (422), g gr d by Exile, (540). Ex. & Br. Mr. Matthew Williams, Bishton Hall, Shifnal.

3rd Prize, £5.—" Earl of Leicester" (171), sire, Royal Chief, (1022) dam by Sir Garnet, (1091), g gr d British Oak, (290).   Ex. & Br. Mr. Joseph Beach, The Hattons, Brewood, Staffordshire.

Reserve Number.—" Double R." (172), sire, Harlescote, (651), dam by Bristol Reserve (144), gr d by Cambrian, (324), g gr d by Exile, (540). Ex. & Br. Mr. Matthew Williams, Bishton Hall, Shifnal.

AGED RAM CLASS.

1st Prize, £15.—" Montford Hero" (160).   Ex. Mr. T. S. Minton.

2nd Prize, £10.—" Sterling " (164).   Ex. Mr. T. J. Mansell.

3rd Prize, £5.—" Royalist" (173), lambed in 1880, sire, Royal Reserve (159), dam by Mansell's No. 16, 1876, (802), gr d by Lord Warden, (777), or Worcester Patron 2nd, (1281). Ex. & Br. Mr. J. Bowen-Jones, Ensdon House, Montford Bridge, R.S.O., Salop.

Reserve Number.—" Prejudice " (174), lambed in 1880, sire, Pride of Bishton, (954), dam by Longbow, (728). Ex. & Br. Mr. J. Lenox Naper, Loughcrew, Oldcastle, Ireland.

𝔍𝔫𝔡𝔢𝔯

TO R. A. S. E. WINNERS.

Lord Taunton (121)
Lord Uffington (56)

Magic (38)
Magnum Bonum (3)
Major (88)
Mansion (46)
Mansion 2nd (55)
Marquis of Bute (91)
Marquis of Lorne (157)
Matthew (79)
Merry Carlisle (156)
Montford Hero (160)

Nobleman (37)
Novelty (41)

Oswestry Champion (137)
Oxford Hero (77)

Paddy (113)
Patentee (4)
Patentee 2nd (20)
Patentee 4th, alias Patentee the Prime (28)
Patentee 5th (42)
Patrician (120)
Pattern (36)
Plymouth Prize (47)
Prejudice (174)
Pride of Hanmer (64)
Prince Regent (169)
Prince Royal (163)
Prince Victor (158)
Proud Salopian (87)

Quality (32)

Raby Duke (108)
Royal Aston (124)
Royal Bristol (138)
Royalist (173)
Royal Ranger (67)
Royal Reserve (159)
Royal Taunton (115)

Second Best (170)
Sheldons' No. 1, 1876 (129)
Sheldons' No. 2, 1878 (143)
Sheldons' No. 1, 1879 (148)
Sir Guy (147)
Sir William (109)
Standard Bearer (80)
Sterling (164)
St. Patrick (29)

Talisman (136)
Taunton Reserve (118)
The Duke (74)
The Ruler (114)
Third Marquis of Bute (132)
Tommy No. 1 (30)
Trojan (130)
True Type (85)

Viceroy (66)
Victor (150)
Volunteer (51)

Worcester Patron (35)

Yardley (145)
Young Duke (140)
Young Quality (52)
Young Sultan (117)

UN-NAMED RAMS.—1, 5, 10, 11, 16, 18, 19, 21, 22, 25, 34, 40, 45 62, 63, 68, 71, 73, 78, 81, 82, 86, 92, 94, 97, 102, 106, 110, 134, 162

3

9

RAMS.

ABREVIATIONS—l. lambed; s. Sire; d. Dam; g.d. Grand Dam; g.g.d. Great Grand Dam; g.g.g.d. Great Great Grand Dam.

N.B.—Name in parenthesis denotes the Breeder of the Ram immediately preceeding it.

"A A." (175) l. in 1881; br. R. Thomas, s AI (176), d by Prince (963), g d by Foreman (568).

"A1," (176) l. in 1878; br. R. Thomas, s Grandeur (133), d by Favourite (553), g d by Young Clifden (1299).

"A1," (177) l. in 1877; br. C. Randell, s British Yeoman (295), d by Our Chief (924).

ACTIVE, (178), l. in 1880; br. R. Thomas, s A1 (176), d by Prince (963), g d by Benthall Chieftain (246).

ADVANCER, (180), l. in 1881; br. A. S. Berry, s Geo. German's No. 14 1880, d a Keeling Ewe.

ALBERT, (181), l. in 1881; br. Crane and Tanner, s Mansell's No. 16, 1876 (802), d by Columbus (128).

ALDERMAN, (182), l. in 1879; br. Mrs. Barrs, s Freeford (Coxon) d by Odstone (918).

ALEXANDER, (183), l. in 1859; br. J. H. Bradburne, s Masfen's No. 15.

ALLSOPP'S No. 2, 1881, (184), l. in 1880; br. Sir H. Allsopp, Bart., s Multum in Parvo (865).

ALPHA, (185), l. in 1880; br. R. Thomas, s A1 (176), d by Prince (963), g d by Benthall Chieftain (246).

ALPINE, (186), l. in 1881; br. Sir H. Allsopp, Bart., s T. J. Mansell's No. 12, 1880.

ANAK, (187), l. in 1880; br. R. H. Masfen, s Clinker (397) d by Capitalist (331).

ANCHOR, (188), l. in 1881; br. W. German, s Cockade (400), d by Marquis of Bath, (822).

ANTIQUE, (189), l. in 1880; br. T. Ryland, s Sir William (1115), d a Crane Ewe.

APOLLO, (190), l. in 1869 ; br. T. Mansell, s Conservative (435), d by Earl of Plymouth (48).

ARISTOCRAT, (191), l. in 1873 ; br. J. W. Minton, s Park Ranger (T. Mansell), d by Lord Clifden (44).

ARISTOCRAT, (192), l. in 1876 ; br. T. Mansell, s Raby Duke (108), d by Landseer (700).

ARLEY, (193), l. in 1880 ; br. T. Ryland, s Brigadier (276), d by Czar (489).

ARTIST, (194), l. in 1874 ; br. T. Mansell, s Landseer (700), d by Conservative (435).

ATTRACTION, (195), l. in 1855 ; br. G. Adney, s Buckskin (Adney), d an Adney Ewe.

AYLESFORD, (196), l. in 1851 ; br. Earl of Aylesford, s an Adney Ram.

BACCHUS, (197), l. in 1878, br. J. Evans, s British Oak (290), d by The Giant (H. Matthews), g d by Mountebank (H, Matthews), g g d by Shrawardine Hero (Crane).

BACCHUS 2ND, (198), l. in 1880 ; br. Thos. Fenn, s Bacchus (197), d by Ensdon Hero (104), g d by Midlothian (842).

BACHELOR, (199) ; br. R. H. Masfen, s Rob Roy (1018), d by Lord Uffington (56), g d by Corinthian, g g d by Mainstay (Masfen).

BACH'S No. 7, 1882, (200), l. in 1881 ; br. F. Bach, s Cockney (401), d by Oswestry Champion (137).

BACH'S No. 18, 1882, (201), l. in 1881 ; br. F. Bach, s Cockney, (401), d by Tarter (1173).

BACH'S RAM, (202), l. in 1853 ; br. late Mr. Bach

BACH'S RAM, (203), l· in 1855 ; br. late Mr. Bach

BACH'S RAM, (204), l. in 1879 ; br. F. Bach

BAKER PASCHA. (205), l. in 1881 ; br. J. Pulley, M.P., s Young Sultan (1328), d by Young Colossus (1301).

BANJO, (206), l. in 1880 ; br. H. J. Sheldon, s British Yeoman (294).

BANKER, (207), l. in 1880 ; br. Lord Chesham, s Son of Beach's, No. 8, 1877, d by Primate (J. Evans).

BANKER, (208), l. in 1881 ; br. G. Allen, s Graphic (631), d a Bostock Ewe.

BANNERMAN, (209). l. in 1880 ; br. Lord Chesham, s Young Colossus (1301), d by Royal Aston (124).

BARON, (210), l. in 1874 ; br. Mrs. S. Beach, s Cardigan (338), d by Duke of Manchester (70).

BARON, (211), l. in 1879 ; br. F. Bach, s Cockney (401), d a· Bach Ewe.

BARON, (212), 1. in 1879 ; br. J. H. Bradburne, s Clinker (397), d a Bradburne Ewe.

BARON, (213), 1. in 1880 ; br. Lord Chesham, s Young Colossus (1301), d by Royal Taunton (115), g d by Lord Kingston (99).

BARON ALKMOND, (214), 1. in 1881 ; br. J. Pulley, M.P., s Young Colossus (1301).

BARON BRISTOL, (215), 1. in 1879 ; br. Crane and Tanner, s Bristol Reserve (144), d by Union Jack (1252), g d by Lord Uffington (56), g g d by Competition (425).

BARONET, (216), 1. in 1881 ; br. J. Evans, s Baron Bristol, (215) d by The Giant (H. Matthews), g d by Mountebank (H. Matthews).

BARON PENDEFORD, (217), 1. in 1879 ; br. R. H. Masfen, s Bristol Prize (142), d by Columbus (128), g d by Caligula (112), g g d by Conqueror (Keeling).

BARON PLASSY, (218), 1. in 1881 ; br. T. Mansell, s Lord Clive (742), d by True Stock (1243).

BARON UFFINGTON, (219), 1. in 1878 ; br. J. Evans, s Royal Taunton (115), d by Grand Duke (620).

BATTUS, (220), 1. in 1881 ; br. J. Evans, s Lord Coxcomb (743), d by Bacchus (197), g d by Hardlines (647).

BEACH'S No. 3, 1873, (221) 1. in 1872 ; br. Mrs. S. Beach, s Cardigan (338), d by Young Gayton (1307).

BEACH'S No. 1, 1875, (222) ; 1. in 1873 ; br. Mrs. S. Beach, s Cardigan (338), d by Cardinal (53).

BEACH'S No. 12, 1875, (223), 1. in 1874 ; br. Mrs. S. Beach, s Peer, d by Quality.

BEACH'S No. 9, 1876, (224), 1. in 1875 ; br. Mrs. Beach, s British Oak (290), d by Briton (296).

BEACH'S No. 19, 1876, (225), 1. in 1875 ; br. Mrs. Beach, s British Oak (290), d by Cardinal (53).

BEACH'S No. 13, 1876, (226), 1. in 1875 ; br. Mrs. S. Beach, s Monarch (863), d by Duke of Manchester (70).

BEACH'S No. 15, 1876, (227), 1. in 1875 ; br. Mrs. S. Beach, s Monarch (863), d by Latimer (H. Smith).

BEACH'S No. 17, 1876 (228), 1. in 1875 ; br. Mrs. S. Beach, s British Oak (290), d by Duke of Manchester (70).

BEACH'S No. 9, 1877, (229), 1. in 1877 ; br. Mrs. S. Beach, s Monarch (863), d by Duke of Manchester (70), g d by Gayton Prince (592).

BEACH'S R.A.S. Ram, (230), 1 in 1879 ; br. J. Beach, s Grand Turk (629), d by Cardinal (53).

BEACH'S No. 10, 1881, (231), 1. in 1880 ; br. J. Beach, s First Choice (561), d by Masterman (828).

BEACH'S No. 12, 1881 (232), l. in 1881 ; br. J. Beach, s First Choice (561), d by British Oak (290).

BEACH'S No. 2, 1882 (233), l. in 1881 ; br. J. Beach, s Royal Chief (1022), d by Manager (795).

BEACH'S No. 3, 1882, (234), l. in 1881 ; br. J. Beach, s Royal Chief (1022), d by Sheldon's No. 2, 1878.

BEACH'S RAM, (235), l. in 1870 ; br. Mrs. S. Beach.

BEACONSFIELD, (236), l. in 1877 ; br. R. H. Masfen, s Columbus (128), d by Commander-in-Chief (416).

BEACONSFIELD, (237), l. in 1878 ; br. W. H. Clare, s Dudmaston (504), d Conductor (90).

BEACONSFIELD, (338), l. in 1879 ; br. T. Mansell, s North Star (913), d by Conservative (435).

BEAUDESERT, (239), l. in 1878 ; br. J. Coxon, s Marquis of Bath (822), d by Confidence (101), g d by Mansion 3rd (818).

BEAUFORT, (240); br. C. Byrd, s Shrawardine Hero (Crane), d by Hampton Hero (84), g d by Oxford Hero (77).

BEDNALL, (241), l. in 1877 ; br. Lord Chesham, s Royal Taunton (115), d by Mansell's No. 8, 1870, g d by Mathew (79).

BEDNALL HERO, (242), l. in 1881 ; br. J. Darling, s Dudmaston Hero (165), d a Keeling Ewe.

BENDIGO, (243), l. in 1875 ; br. C. Timmis, s Earl of Evesham (520), d by Model Patentee.

BENEDICT, (244), l. in 1880 ; br. Lord Chesham, s Royal Aston (124), d by British Tar (292).

BENEDICT, (245), l. in 1879 ; br. R. H. Masfen, s B. B., (T. J. Mansell), d by Caligula (112), g d by Corsair (57), g g d by Corinthian (Coxon).

BENTHALL CHIEFTAIN, (246), l. in 1869 ; br. J. & E. Crane, s Chieftain (384), d a Crane Ewe.

BENTICK, (247), l. in 1879 ; br. R. Thomas, s Prince (963), d by Calcot Chieftain (319), g d by Young Clifden (1299).

BERKELEY, (248), l. in 1878 ; br. R. H. Masfen, s Bristol Prize (142), d by Commander-in-Chief (416), g d by Corsair (57).

BERRINGTON HERO, (249), l. in 1872 ; br. R. V. C. Groves, s Volunteer 2nd (1261), d a Meire Ewe.

BERWICK, (250), l. in 1881 ; br. T. S. Minton, s Montford Hero (160), d a Minton Ewe.

BERWICK HERO, (251), l. in 1882 ; br. T. S. Minton, s His Lordship 2nd (670), d by Calcot (317).

BETTON, (252), l. in 1874 ; br. Messrs. Minton, s Son of Little Lord (1139). d a Minton Ewe.

BIG BACK, (253), l. in 1880; br. J. Beach, s First Choice (561), d by British Oak (290).

BIG CORPORAL, (254), l. in 1879; br. C. Randell, s Colossal (410), d by Corporal (444).

BIG GUN, (255), l. in 1879; br. H. Rogers, s a Sheldon Ram, d a Wolgarstone Ewe.

BIG REDNAL, (256), l. in 1873; br. E. Meredith, s Long Back d a Meredith Ewe.

BIRKENHEAD, (257), l in 1880; br. Crane & Tanner, s Columbus (128), d by Chivalry (387).

BISHTON, (258), l. in 1880; br. M. Williams, Junr., s Pride of Bishton (954), d by Artist (194).

BLACKALL, (259), l. in 1881; br. Lord Chesham, s Son of Royal Aston, d by Son of Mansell's No. 6.

BLACK KNIGHT, (260), l. in 1881; br. R. Thomas, s Cœur de Lion (404), d by Favourite (553), g d by Castle Warden (342).

BLOOD ROYAL, (261), l. in 1879; br. F. Bach, s Sir Joseph (1101), d by Oswestry Champion (137).

BLUE JACKET, (262), l in 1880; br. Lord Chesham, s British Tar (292), d a Chesham Ewe.

BOREAS, (263), l. in 1877; br. Mrs. S. Beach, s British Oak (290), d by Cardigan (338).

BOUNCER, (264), l in 1876; br. J. Coxon, s Champion (351), d by Confidence (101).

BOURTON, (265), l. in 1881; br. E. Instone, s a Chesham Ram (134), d by a T. J. Mansell Ram.

BRABAZON, (266), l in 1881; br. T. Nock, s Jumbo, d by True Light (1242).

BRADBURNE'S No. 26, 1875 (267), l. in 1874; br. J. H. Bradburne.

BRAILES, (268), l in 1879; br. H. J. Sheldon, s Kilburn Commended.

BREAD WINNER, (269), l, in 1881; br. T. Mansell, s Lord Clive (742), d by Multum (882).

BREASTPLATE, (270), l. in 1878; br. J. Evans, s British Oak (290), d by Hardlines (647), g d by Standard Bearer (80).

BRERETON BOY, (271), l. in 1879; br. The Earl of Shrewsbury, s Forton Hero (575), d by Hercules (662), g d by Curzon's No. 32, 1863, g g d by Symmetry.

BREWOOD, (272), l. in 1880; br. J. Beach, s First Choice (561), d by Cardigan (338).

c

BREWOOD CHIEF, (273), l. in 1881 ; br. J. Beach, s. Royal Chief (1022), d by Masterman (830).

BRIDGNORTH, (274), l. in 1879 ; br. T. J. Mansell, s His Lordship (669), d by May Duke (837).

BRIGADIER, (275), l. in 1876 ; br. John Evans, s British Oak (290), d an Evans Ewe.

BRIGADIER, (276), l. in 1878 ; br. C. Byrd, s Birmingham Reserve (126), d a Crane Ewe.

BRIGAND, (277), l. in 1876 ; br. J. Evans, s British Oak (290), d by Claudius (103), g d by Cardinal (53).

BRILLIANT, (278), l. in 1880 ; br. J. Evans, s Bristol Reserve (44), d by Royal Taunton (115).

BRINKS, (279), l. in 1879 ; br. T. Fenn, s Brilliant (Masfen), d by Ensdon Hero (104), g d by Midlothian (842).

BRISBANE, (280), l. in 1879 ; br. T. S. Minton, s His Lordship (669), d by Bedford Hero (111).

BRISTOL BEAU, (281), l. in 1881 ; br. J. Evans, s Bristol Reserve (144), d by Royal Taunton (115), g d by Grand Duke (620).

BRISTOL CHIEFTAIN, (282), l. in 1882 ; br. J. Evans, s Baron Bristol (215), d by Grand Duke (620), g d by Chieftain (384), g g d by Nonpareil (908).

BRISTOL DUKE, (283), l. in 1879 ; br. J. Evans, s Bristol Reserve (144), d by British Oak (290), g d by Union Jack (1252).

BRISTOL PET, (284), l. in 1881 ; br. J. Evans, s Baron Bristol (215), d by The Giant (H. Matthews), g d by Mountebank (H. Matthews), g g d a Matthews Ewe.

BRISTOL PRINCE, (285), l. in 1879 ; br. Crane & Tanner, s Bristol Reserve (144), d by Corsair (57).

BRISTOL PRINCE, (286), l. in 1881 ; br. J. Evans, s Bristol Reserve (144), d by May Duke (837), g d by Union Jack (1252).

BRITANNICUS, (287), l. in 1878 ; br. J. Evans, s British Oak (290), d Union Jack (1252), g d by Cardinal (53).

BRITISH CHIEFTAIN, (288), l. in 1878 ; br. C. Randell, s Our Chief (924), d by British Yeoman (295)

BRITISH FLAG, (289), l. in 1881 ; br. Sir H. H. Vivian, Bart., s Nightingale's No. 2, 1830 (896), d an Amies Ewe.

BRITISH OAK, (290), l. in 1873 ; br. J. Evans, s Grand Duke (620), d by Hardlines (647), g d by Competition (425), g g d by Emperor (525).

BRITISH PRINCE, (291), l. in 1876 ; br. J. Evans, s British Oak (290), d by Broadguage (89), g d by Young Emperor (1305).

BRITISH TAR, (292), l. in 1876 ; br. J. Evans, s British Oak (290), d by Cardinal (53), g d by Pride of Pitchford (961), g g d by Young Emperor (1305).

BRITISH TAR, (293), l. in 1879 ; br. Viscount Falmouth, s Masterman (828), d by Harry (656).

BRITISH YEOMAN, (294), l. in 1877 ; br. E. Meredith, s The Tory (1208), d by Montford (869).

BRITISH YEOMAN, (295), l. in 1875 ; br. Mrs. S. Beach, s British Oak (290), d by Duke of Manchester (70).

BRITON, (296), l. in 1868 ; br. the late J. Beach, s Cardinal (53), d by Gayton Prince (592).

BROAD BACK, (297), l. in 1881 ; br. Miss Rose, s Nock's No. 6, d by Fat Back, g d by Lord Hereford (Lord Aylesford).

BROMFIELD, (298), l. in 1881 ; br. T. Fenn, s Mansell's No. 8, d by Ensdon Hero (104), g d by Midlothian (842).

BROMLEY'S No. 13, 1879 (299), l. in 1878 ; br. R. Bromley, s Lord Harlech (135), d by Young Hardlines (R. Bromley).

BROMLEY'S RAM, (300), l. in 1879 ; br. R. Bromley, s Wonder (1277), d by Hardlines (647).

BRONTES (301), l. in 1881 ; br. J. Evans, s Baron Bristol (215), d by Royal Taunton (115), g d by May Duke (837).

BROTHER TO EARL OF WARWICK, (302), l. in 1857 ; br. H. J. Sheldon, s Attraction (195). d a Sheldon Ewe.

BROTHER TO LORD LIVERPOOL, (303), l. in 1877 ; br. Capt. H. Townshend, s Example (127), d a Masfen Ewe.

BRUCE, (304), l. in 1871 ; br. R. H. Masfen, s Rob Roy (1018), d by Monarch (Masfen), g d by Brother to Gratitude (Masfen), g g d by Lord Harley (Adney).

BRUTUS, (306), l. in 1876 ; br. J. Evans, s British Oak (290), d by Royalty (1029), g d by Premier (947).

BUCKSKIN, (307), l. in 1870 ; br. G. Allen, s Fat Backed Patentee.

BULWARK, (308), l. in 1879 ; br. R. H. Masfen, s Bristol Prize (142), d by Columbus (128), g d by Caligula (112), g g d by True Type (85).

BUMPER, (309), l. in 1880 ; br. J. Beach, s First Choice (561), d by Cardigan (338).

BURDETT COUTTS, (310), l. in 1877 ; br. Mrs. S. Beach, s Colossus (411), d by The Knight (1160), g d by Latimer (Smith).

BURWAY CRANE, (311), l. in 1879 ; br. V. E. Nightingale, s Carouser (341), d by Claudius (103), or Caligula (112).

BURWAY PIPPIN, (312), l. in 1880; br. V. E. Nightingale, s Downton Pippin (503), d by Claudius (102), or Caligula (112)

BYRD'S No. 4, 1873 (313), l. in 1872; br. C. Byrd, s Oxford Hero (77), d by Blood Royal (83).

BRYD'S No. 18, 1875 (314), l. in 1874; br. C. Byrd, s The General (1186), d by Manager (794).

BYRD'S No. 36, 1880 (315), l. in 1879; br. C. Bryd, s Victor (150), d by The General (1186), g d by Earl of Chester (519).

BYRON, (316), l. in 1876; br. Mrs. H. Smith.

CALCOT, (317), l. in 1870; br. Joseph Crane, s Crosswood Hero (65), d by Big Plymouth (49)

CALCOT 2nd, (318), l. in 1874; br. J. Bowen-Jones, s Conserva-ative Calcot (436) d by Lord Warden (777).

CALCOT CHIEFTAIN, (319), l. in 1868; br. J. Crane, s Chieftain (384), d by Celebrity (6).

CALDER, (320), l. in 1880; br. R. Jefferson, s Lothair (779), d a Mrs. Smith Ewe.

CALDICOTE, (321), l. in 1877; br. Capt. H. Townshend, s Example (127), d a Townshend Ewe.

CALIGULA 2nd, (322), l. in 1876; br. E. Crane, s Caligula (112), d by Cambrian (324). g d by Plymouth Prize (47).

CALIGULA 2nd, (323), l. in 1879; br. Crane & Tanner, s Caligula (112), d by Celebrity (6).

CAMBRIAN (324), l. in 1870; br. J. & E. Crane, s Crosswood Hero (65), d by Duke of Newcastle (39).

CANDIDATE, (325), l. in 1881; br. M. Barrs, s Lord Oxon (167), d by Corporal (443).

CANNOCK CHIEF, (326), l. in 1876; br. R. H. Masfen, s Hero (F. Byrd), d by Norton (Masfen), g d by Commander-in-Chief (416), g g d by Cardinal (53).

CANTAB, (327), l. in 1878; br. R. H. Masfen, s Clinker (397), d by Capitalist (331), g d by Norton (Masfen), g g d by Corsair (57), g g g d by Mainstay (Masfen), g g g g d by Lord Harley (Adney), g g g g g d by Lady Jane Grey's Son (Masfen).

CANTERBURY 2nd, (328), s Canterbury Patentee (15).

CANTERBURY 3rd, (329), s Canterbury 2nd (328).

CAPITAL, (330), l. in 1879; br. C. Randell, s Colossal (410), d by Typical (1249).

CAPITALIST, (331), l. in 1872; br. J. Coxon, s Captivator (76), d by Commander (415).

CAPTAIN, (332), l. in 1876 ; br. C. Byrd, s Captor (334), d by Major (88).

CAPTAIN THOMAS, (333), l. in 1877 ; br. Mrs. S. Beach, s Colossus (411), d by Reflection (998), g d by Cardinal (53).

CAPTOR, (334), br. Mrs. S. Beach, s. Cardigan (338).

CARACTACUS, (335), l. in 1865 ; br. J. & E. Crane, s Chieftain (384), d by Celebrity (6).

CARADOC, (336), l. in 1859 ; br. J. & E. Crane, s Celebrity (6), d a Crane Ewe.

CARDIFF, (337), l. in 1879 ; br. Crane & Tanner, s Columbus (128), d by Chivalry (387).

CARDIGAN, (338), l. in 1872 ; br. R. H. Masfen, s Commander-in-Chief (416), d by Cardinal (53).

CARDINAL, (339), l. in 1879 ; br. Lord Chesham, s Son of Mansell's No. 6, d by Son of No. 12, g d by Patron.

CARDINAL, 2nd, (340), l. in 1874 ; br. R. Barber, s Son of Cardinal (1126), d a Barber Ewe.

CAROUSER, (341), br. J. Evans (34) ; s British Tar (292), d by Union Jack (1252).

CASTLE WARDEN, (342), l. in 1869 ; br. P. W. Bowen, s Lord Warden (777), d by Courtier (453).

CAVALIER, (343), l. in 1874 br. J. Evans, s Claudius (103), d by Cardinal (53), g d by Volunteer (51), g g d by Emperor (525)

CAVALIER, (344), l. in 1881 ; br. T. Mansell, s Multum (882), d by Raby Duke (108).

CELERITY, (345), l. in 1877 ; br. Crane & Tanner, s son of Chivalry (1128), d by Claudius (103).

CHADBURY, (346), l. in 1881 ; br. C. Randell, s Quarryman (986), d by Julius Cæsar (692), g d by Conservative.

CHALLENGE, (347), l. in 1880 ; br. T. Mansell, s Multum in Parvo (885), d by Calcot (317).

CHALLENGER, (348), l. in 1875 ; br. J. Evans, s Union Jack (1252), d by Cardinal (53).

CHALLENGER, (349), br. C. Bryd, s Chivalry (387), d by Legatee (95), g d by Guardian (642).

CHALLENGER, (350) l. in 1880 ; br. Mrs. M. Barrs, s Czar (488), d a Walker Ewe.

CHAMPION, (351), l in 1873 ; br. J. Coxon, s Stamina (1157), d by Commander (415).

CHAMPION, (352), l. in 1879 ; br. Lord Chesham, s Beach's No. 8, 1877, d by Royal Aston (124), g d by Lord Vincent (Chesham).

38

CHÅNCELLOR, (353), l. in 1868; br. J. Coxon, s Commander (415), d by Cavalier (Thornton), g d by Patent (931).

CHANCELLOR, (354), l. in 1879; br. Lord Chesham, s Son of Colossus, d by Royal Taunton (115), g d by (Lord Kingston) (99).

CHANCELLOR, (355), l. in 1879; br. C. Byrd, s Victor (150), d by The General (1186), g d by Earl of Chester (519).

CHANDOS, (356), l. in 1879; br. R. H. Masfen, s Clinker (397), d by Lord Harlech (135), g d by Cardigan (338), g g d by Corinthian (Coxon).

CHARACTER, (357), l. in 1878; br. R. Edwards, s Caligula 2nd (322).

CHARLES 2nd, (359), l. in 1883; br. Viscount Falmouth, s Hattons British Oak (657), d a Chesham Ewe.

CHATSWORTH, (360), l. in 1881; br. J. Evans, s Lord Coxcomb (743), d by Union Jack (1252), g d by Grand Duke (620).

CHESHAM, (361), l. in 1879; br. R. H. Masfen, s Clinker (397), d by Capitalist (331), g d by True Type (85), g g d by Corsair (57).

CHESHAM 2nd, (362), l. in 1879; br. R. Loder, s Lord Chesham, d a Chesham Ewe.

CHESHAM'S No. 10, 1871 (363), l. in 1870; br. Lord Chesham.

CHESHAM'S No. 19, (364), l. in 1872; br. Lord Chesham, s Son of Duke of Manchester.

CHESHAM S No. 9, 1875 (365), l. in 1874; br. Lord Chesham, s Chesham's No. 12, d by Matthew (79).

CHESHAM'S No. 10, 1876 (366), l. in 1875; br. Lord Chesham, s Young Harrogate (Chesham), d by Latimer (Smith).

CHESHAM'S No. 15, 1876 (367), l in 1875; br. Lord Chesham, s Son of Pattern (Chesham), d by T. Mansell's No. 8, g d a Chesham Ewe.

CHESHAM'S No. 19, 1876 (368), l. in 1875; br. Lord Chesham, s Marquis of Bute (91), d by Duke of Manchester (70).

CHESHAM'S No. 20, 1876 (369), l. in 1875; br. Lord Chesham, s Young Harrogate (Chesham), d by Masterman (828).

CHESHAM'S No. 8, 1881 (370), l. in 1880; br. Lord Chesham, s a Chesham Ram.

CHESHAM'S No. 26, 1881 (371), l. in 1880; br. Lord Chesham, s Son of Young Colossus (Chesham), d by Royal Taunton (115), g d by Lord Kingston (99).

39

CHESHAM'S No. 44, 1881 (372), l. in 1880; br. Lord Chesham, s Son of Young Colossus (Chesham), d by Royal Taunton (115).

CHESHAM'S No. 1, 1882 (373), l. in 1881; br. Lord Chesham, s Dudmaston (506), d by Beach's No. 1, 1877.

CHESHAM'S No. 2, 1882 (374), l. in 1881; br. Lord Chesham, s Son of Royal Aston (Chesham), d by Mrs. Beach's No. 1, 1877.

CHESHAM'S No. 5, 1882 (375), l. in 1881; br. Lord Chesham, s Dudmaston (506), d by Royal Taunton (115).

CHESHAM'S No. 8, 1882 (376), l. in 1881; br. Lord Chesham, s Dudmaston (506), d by Mrs. Beach's No. 1.

CHESHAM'S No. 14, 1882, (377), l. in 1881; br. Lord Chesham, s Son of Royal Aston (Chesham), d by Mrs. Beach's No. 1.

LORD CHESHAM'S NOCK, (378), l. in 1872; br. Lord Chesham s Son of Mansell's No. 8 (Chesham), d a R.A.S.E. Prize Ewe.

CHESHAM RAM, (379), l. in 1877; br. Lord Chesham.

CHESHAM'S RAM, (380); br. Lord Chesham, s Royal Aston (124), d an Evans' Ewe.

CHESTER, (381), l. in 1857; br. Mr. Adney, s Buckskin (Adney).

CHIEF, (382), l. in 1877; br. R. H. Masfen, s Columbus (128), d by Corsair (57), g d by Corinthian (Coxon), g g d by Brother to Gratitude (Masfen).

CHIEF, (383), l. in 1876; br. Crane & Tanner, s Chivalry (387), d by Claudius (103).

CHIEFTAIN, (384), l. in 1862; br. J. & E. Crane, s Caradoc (336), d by Tern (1176).

CHIEFTAIN, (385), l. in 1879; br. J. Coxon, s Sir Robert (1108).

CHIEFTAIN, (386), l. in 1873; br. C. Randell, s Our Chief (924), d by Lord Uffington (56) or Cardinal (53).

CHIVALRY, (387), l. in 1872; br. E. Crane, s Cambrian (324), d d by Lord Uffington (56).

CHRISTY, (388), l. in 1875; br. E. Crane, s Landseer (700), d by Claudius (103).

CICERO, (389). l. in 1870; br. R. Thomas, s Mansell's No. 6, 1868, (797), d by The Ruler (1204).

CIRCUIT, (390), l. in 1881; br. Lord Chesham, s Beach's No. 8, d by M. Williams No. 15, 1877.

CITY MEMBER, (391), l. in 1878; br. T. J. Mansell, s County Member (452), d by Calcot (317).

CLAIMANT, (392), l. in 1872; br. J. Coxon, s Stamina (1157), d
by Commander (415).

CLANSMAN, (393), l. in 1881; br. J. Evans, s Lord Coxcomb
(743), d by Grand Duke (620), g d by Claudius (103).

CLAUDE DUVAL (394), l. in 1874; br. E. Crane, s Claudius
(103), d by Lord Uffington (56).

CLAUDIUS 2nd, (395), l. in 1876; br. J. Bowen-Jones, s Claudius
(103), d by Fosse Duke (576).

CLAUDIUS 3rd, (396), l. in 1877; br. Mr. Walker, s Claudius
2nd (Walker).

CLINKER, (397), l. in 1876; br. R. H. Masfen, s Columbus (128),
d by Conqueror, g d by Corsair (57), g g d by Corinthian
(Coxon), g g g d by Old Crop (Masfen), g g g g d by Brother
to Gratitude (Masfen).

CLINKER, (398), l. in 1877; br. Crane & Tanner, s Birmingham
Reserve (126), d by Chivalry (387).

CLIPSTONE, (399), l. in 1865; br. J. Coxon, s Duke of Newcastle
(39), d by Patent (931).

COCKADE, (400), l. in 1879; br. Crane & Tanner, s Columbus
(128), d by Chivalry (387).

COCKNEY, (401), l. in 1878; br. Crane & Tanner, s Coxcomb
(457), d by Chivalry (387).

COCK OF THE NORTH, (402), l. in 1879; br. T. J. Mansell.

COCKSWAIN, (403), l. in 1881; br. J. Coxon, s Cockade (400),
d by Champion (351).

CŒUR-DE LION, (404), l. in 1879; br. R. H. Masfen, s Columbus
(128), d by Caligula (112), g d by True Type (85).

COHESION, (405), l. in 1869; br. J. Coxon, s Commander (415),
d by Novelty (41).

COLDSTREAM, (406), l. in 1876; br. Viscount Falmouth, s
Secundus (1049), d by Lord Paramount (772).

COLLINGWOOD, (407), l. in 1879; br. C. Randell, s Nelson (889),
d by Corporal (444).

COLONEL, (408), l. in 1878; br. by Lord Chesham, s Royal
Aston (124), d by No. 12, g d by Latimer (H. Smith).

COLONEL, (409), l. in 1880; br. J. Pulley, s Sir Roger (1112).

COLOSSAL, (410), l. in 1877; br. Mrs. S. Beach, s Colossus (411),
d by Caractacus (335), g d by Nock's No 4, 1868.

COLOSSUS, (411), l. in 1874; br. Mrs. Beach, s Cardigan (338),
d by Gayton Prince (592).

COLOSSUS 3rd, (412), l. in 1880; br. J. Pulley, s Young Colossus
(1301), d by Young Sultan (1328).

COLSTON, (413), l. in 1879; br. R. M. Knowles, s Salisbury (1038), d a Keeling Ewe.

COMET, (414), l. in 1880; br. Viscount Falmouth, s Goliath (612), d a Falmouth Ewe.

COMMANDER, (415), l. in 1866; br. A. H. Minor, s Volunteer (51), d by Patentee the Prime (28).

COMMANDER-IN-CHIEF, (416), l. in 1868; br. J. Coxon, s Commander (51), d by Novelty (41).

COMMERCE, (417), l. in 1880; br. Crane & Tanner, s Columbus (128), d by Dudmaston (504).

COMMODORE, (418), l. in 1869; br. W. German, s Commander (415), d by Patent (931).

COMMODORE, (419); br. Lord Willoughby de Broke.

COMMODORE, (420), l. in 1882; br. C. Randell, s Collingwood (407), d by Jack Tar (683).

COMMISSIONER, (421). l. in 1877; br. H. J. Sheldon, s Taunton Reserve (118), d a Sheldon Ewe.

CO-MONUMENT, (422), l. in 1871; br. T. Mansell, s Son of Conservative (T. Mansell), d by Marquis (820).

CO-MONUMENT, (423), l. in 1870; br. T. Mansell, s Son of Conservative (T. Mansell), d by Mansion 2nd (55), g d a Mansell Ewe.

COMPASS, (424), l. in 1880; br. J. Coxon, s City Member (391), d by Sir Robert (1108), g d by Confidence (101).

COMPETITION, (425), l. in 1861; br. S. Byrd, s Quality (J. Evans), d by Patentee (4), g d an Old Farmer Ewe.

COMUS, (426), l. in 1879; br. Lord Chesham, s Son of Colossus (Chesham), d by Son of Marquis of Bute (Chesham), g d by Lord Kenyon.

CONCORD, (427), l. in 1879; br. T. Mansell, s County Member (452), d by Calcot (317).

CONDOR, (428), l. in 1880; br. Crane & Tanner, s Yardley (1283), d by Dudmaston (504).

CONFEDERATE, (429), l. in 1871; br. R. H. Masfen, s Commander-in-Chief (416), d by Novelty (41).

CONGRESS, (430), l. in 1878; br. J. Coxon, s Sir Robert (1108).

CONGREVE, (431), l. in 1879; br. C. R. Keeling, s Shipston (1074), d by The Count (Keeling).

CONNAUGHT, (432), l. in 1881; br. Lord Chesham, s Son of Colossus (Chesham), d by Primate (J. Evans).

CONQUEROR (433), l. in 1874; br. J. H. Bradburne, s Commandant (Keeling), d by Crosswood Hero (65).

CONQUEROR, (434), l. in 1881 ; br. T. F. Cheatle, s Warrior (1265), d by Beaconsfield (236).

CONSERVATIVE, (435), l. in 1864 ; br. Mrs. Wadlow, s P. W. Bowen's No. 11, 1862 ; d a Mrs. Wadlow Ewe.

CONSERVATIVE CALCOT, (436), l. in 1872 ; br. T. Mansell, s Calcott (317), d by Conservative (435).

CONSERVER, (437), l. in 1873 ; br. J. Coxon, s Confidence (101), d by Preserver (950).

CONSOLATION, (438), l. in 1871 ; br. J. Coxon, s Freeman (C. Byrd), d by Compton Buck (S. Byrd).

CONSTANTINE, (439) ; br. Lord Chesham, s Royal Grand Duke (1025).

CONWAY, (440), l. in 1872 ; br. J. & E. Crane, s Cambrian (324), d by Lord Uffington (56).

COPPERPLATE, (441), l. in 1872 ; br. C. R. Keeling, s True Type (85), d by Lord Uffington (56), g d by Patentee the Prime (28).

CORMORANT, (442), l. in 1875 ; br. E. Crane, s Chivalry (387), d by Buckskin (307), g d by Corsair (57).

CORPORAL, (443), l. in 1871 ; br. W. H. L. Clare, s Conductor (90), d by President.

CORPORAL, (444), l. in 1873 ; br. C. Randell, s Field Marshall (559), d by Maddox's Ram.

COSSACK, (445), l. in 1875 ; br. R. H. Masfen, s Columbus (128,) d by Norton (R. H. Masfen), g d by Corsair (57), g g d by Mainstay (Masfen), g g g d by Lord Harley (Adney), g g g g d by Lady Jane Grey's Son, g g g g g d by Stratton (Stratton).

COSSACK (446), l. in 1874 ; br. R. H. Masfen, s His Majesty (T. J. Mansell), d by Commander-in-Chief (116).

COSTON, (447), l. in 1879 ; br. R. Thomas, s Prince (963), d by Calcot Chieftain (319), g d by The Ruler (1204).

COUNT, (448), l. in 1871 ; br. R. Thomas, s Marquis (820), d by a Crane Ram.

COUNT, (449), l. in 1876 ; br. G. German, s Colossus (411), d by Lord Haughton (751).

COUNT, (450), l. in 1876, br. R. H. Masfen, s Columbus (128), d by True Type (25), g d by Corsair (57), g g d by Turpin (Masfen), g g g d by Mainstay (Masfen), g g g g d by Black Prince (Adney).

COUNTRYMAN (451), l. in 1880 ; br. T. S. Minton, s Royal Reserve (159), d by Severus (T. J. Mansell).

COUNTY MEMBER, (452), l. in 1876 ; br. J. Coxon, s Champion (351), d by Confidence (101), g d by Preserver (950).

COURTIER, (453), br. J. & E. Crane, s Celebrity (6).

COURTIER, (454), l. in 1874; br. W. German, s Confidence (101), d by Chancellor (69).

COURTIER, (455), l. in 1877; br. J. Coxon, s Magnet (788), d by Captivator (76).

COVENTRY, (456), l. in 1863; br. H. J. Sheldon, s Lord Astley (735), d a Sheldon Ewe.

COXCOMB, (457), l. in 1875; br. E. Crane, s Caligula (112), d by Corsair (57).

COXON, (458), l in 1880; br. Capt. J. B. Haydock, s Coxon's No. 8, d by Count (450).

COXON'S CRITIC, (459), l. in 1872; br. Messrs. Coxon.

COXON'S No. 6, 1876 (460), l. in 1875; br. J. Coxon, s Champion (351), d by Commander (415), g d by Novelty (41).

COXON'S No. 14, 1878 (461), l. in 1877; br. J. Coxon, s Courtier (455), d by Confidence (101), g d by Apollo (T. Mansell).

COXON'S No. 4, 1880 (462), l. in 1879; br. J. Coxon, s Masterpiece (832), d by Champion (351).

COXON'S No. 20, 1882 (463), l. in 1881; br. J. Coxon, s Beaudesert (239), d by Captivator (76).

CRAFTY, (464), l. in 1859; br. J. & E. Crane, s Celebrity (6), d by Tern (1176).

CRANE'S No. 11, 1868. (465), l. in 1867; br. J. & E. Crane, s Sheet Anchor (1060), d by Celebrity (6).

CRANE'S No. 17, 1868 (466), l. in 1867; br. J. & E. Crane, s Big Plymouth (49), d by Tern (1176).

CRANE'S No. 8, 1872 (467), l. in 1871; br. E. Crane, s Mansell's No. 1 (T. Mansell), d by Lord Uffington (56).

CRANE'S No. 11, 1873 (468), l. in 1872: br. E. Crane, s Claudius (103), d Crosswood Hero (65).

CRANE'S No. 3, 1875 (469), l. in 1874; br. E. Crane s Claudius (103), d by Lord Uffington (56).

CRANE'S No. 6, 1876 (470), l. in 1875; br. E. Crane, s Claudius 103, d by Crosswood Hero (65).

CRANE'S No. 12, 1876 (471), l. in 1875; br. E. Crane.

CRANE & TANNER'S No. 7, 1877 (472), l. in 1876; br. Crane and Tanner, s Caligula (112), d by Claudius (103).

CRANE & TANNER'S, No. 2, 1879 (473), l. in 1878; br. Crane and Tanner, s Columbus (128), d by Caligula (112).

CRANE & TANNER'S No. 8, 1879 (474), l. in 1878; br. Crane and Tanner, s Dudmaston (504), d by Caligula (112).

CRANE & TANNER'S No. 16, 1879 (475), l. in 1878, br. Crane and Tanner, s Dudmaston (504), d by Chivalry (387).

CRANE & TANNER'S No. 18, 1879 (476), l. in 1877 ; br. Crane and Tanner, s Dudmaston (504), d by Claude Duval (394).

CRANE & TANNER'S No. 5, 1881 (477), l. in 1880 ; br. Crane and Tanner, s Columbus (128), d by Claudius (103).

CRANE & TANNER'S No. 4, 1882 (478), l. in 1881 ; br. Crane and Tanner, s Bristol Prince (285), d by Claudius (103).

CRANE & TANNER'S No. 6, 1882 (479), l. in 1881 ; br. Crane and Tanner, s Mansell's No. 16, 1876 (802), d by Claudius (103).

CRANE & TANNER'S No. 7, 1882 (480), l. in 1881 ; br. Crane and Tanner, s Bristol Prince (285), d by Columbus (128).

CROMWELL, (481), l. in 1881 ; br. Mrs. M. Barrs, s Lord Oxon (167), d by Czar (488).

CRONKHILL, (482), l. in 1869 ; br. Rev. Noel Hill.

CROWN PRINCE, (483), l. in 1870 ; br. W. German, s Conductor (90), d by Novelty (41).

CYNIC, (484), l. in 1877 ; br. R. H. Masfen, s Columbus (128), d by True Type (85), g d by Commander-in-Chief (416).

CYNOSURE, (485), l. in 1877 ; br. R. H. Masfen, s Columbus (128), d by Cardigan (338), g d by Commander-in-Chief (416), g g d by Norton (Masfen).

CYPRUS, (486), l. in 1877 ; br. J. W. Minton, s Bedford Hero (111), d a Minton Ewe.

CYRUS, (487), l. in 1879 ; br. Crane & Tanner, s Dudmaston (504), d by Claudius (103).

CZAR, (488), l. in 1876 ; br. G. German, s Colossus (411), d by Confidence (101).

CZAR, (489), l. in 1876 ; br. R. H. Masfen, s Columbus (128), d by Norton (Masfen).

DANDY, (490), l. in 1880 ; br. J. Bowen-Jones, s Dudmaston (21), 1878, d by Claudius (103).

DAWES' CLINKER, (491,) l. in 1880 ; br. W. M. Dawes, s Oney (922), d by Farmer's No. 30, 1877.

D.C., (492), l. in 1882 ; br. C. Randell, s D.D. (493), d by Colossal (410).

D.D., (493), l. in 1879 ; br. Crane & Tanner, s Dudmaston (504), d by Claudius (103).

DE BROKE, (494), l. in 1863 ; br. Lady Willoughby de Broke, s Sir Harry (1098), d by Duke of Kent (13).

DEBTOR, (495), l. in 1877; br. J. Coxon, s Champion (351), d by Confidence (101), g d by Commander (415).

DEFIANCE, (496), l. in 1871; br. C. Byrd, s Young Napier (1319), d by Guardian (642).

DEFIANCE, (497), l. in 1874; br. J. Evans, s Union Jack (1252), d by Cardinal (53), g d by Humphrey Davy (676).

DIGNITY, (498), l. in 1881; br. J. A. Barrs, s Lord Oxon (167), d by Lord Odstone (766), g d by Claudius (103).

DISAPPOINTMENT, (499), l. in 1882; br. T. J. Mansell, s Warwick (1268), d by County Member (452), g d by May Duke (837).

DOUBLE B., (500), l. in 1873; br. E. Crane, s Claudius (103).

DOUBLE X, (501), l. in 1877; br. T. J. Mansell, s Double B (500), d by True Stock (1243), g d by Calcot (317).

DOWNTON PIPPIN, (502), l. in 1868; br. J. Evans, s Nonpareil (908), d by Competition (425).

DOWNTON PIPPIN, (503), l. in 1878; br. T. Fenn, s My Lord (T. Fenn), d by Bruce (304), g d by Lord Kenyon.

DUDMASTON, (504), l. in 1875, br. T. J. Mansell, s Double B. (500), d by Pattern (933).

DUDMASTON, (505), l. in 1879; br. T. J. Mansell, s Pride of Montford (959), d by Double B. (500), g d by True Stock (1243).

DUDMASTON, (506), l. in 1879; br. T. J. Mansell, s Pride of Montford (959), d by True Stock (1243), g d by Son of Conservative (1134).

DUDMASTON No. 21, 1878 (507), l. in 1877; br. T. J. Mansell, s May Duke (837), d by Landseer (700).

DUKE GOLIAH, (508), l. in 1879; br. J. Harding, s His Lordship (669), d by Sheldon's No. 3, 1876 (1063), g d by Marquis (820).

DUKE OF CAMBRIDGE, (509), l. in 1872; br. C. R. Keeling, s Commander-in-Chief (416), d by Corsair (57), g d by Grindle (Keeling).

DUKE OF CORNWALL, (510), l. in 1875; br. J. Evans, s Grand Duke (620), d by Royalty (1029), g d by Premier (947).

DUKE OF EDINBORO' (511), l. in 1868; br. C. W. Hamilton, s The Duke, d a Hamilton Ewe.

DUKE OF HATTON'S, (512), l. in 1876; br. Mrs. S. Beach, s Admiral (Andrews), d by Cardigan (338), g d by Young Gayton (1307).

DUKE OF LEINSTER, (513), l. in 1865; br. C. W. Hamilton, s The Duke, d by Juvenile (8).

DUKE OF MARLBORO, (514), l. in 1879; br. J. L. Naper, s Cossack (445), d by Conway (440).

DUKE OF WELLINGTON, (515), l. in 1874; br. M. Williams, s Co-Monument (422), d by Lord Napier (764).

DUNESKE, (516), l. in 1879; br. F. Gretton, s Sheldon's No. 4, 1878, d by Prince Imperial (966), g d by Latimer (H. Smith).

EARLDOM, (517), l. in 1880; br. J. Evans, s Royal Taunton, (115), d by Cavalier (343).

EARL OF BEDFORD, (518), l. in 1873; br. Mrs. S. Beach, s Cardigan (338), d by Rejected.

EARL OF CHESTER, (519), l. in 1872, s Oxford Hero (77), d by Guardian (642).

EARL OF EVESHAM, (520), l. in 1872; Mrs. H. Smith, s Lord Grey (H. Smith), d a Smith Ewe.

EARL OF SHREWSBURY, (521), l. in 1878; br. M. Williams, s Bristol Reserve (144), d by Young Hardlines.

ECONOMY, (522), l. in 1880; br. J. Evans, s Bristol Reserve (144), d by The Giant (H. Matthews), g d by Mountebank (H. Matthews).

ELLESMERE, (523), l. in 1873; br. E. Meredith, s Montford (869), d a Meredith Ewe.

EMIGRANT, (524), l. in 1864; br. J. Evans, s Pride of Pitchford (961), d by Emperor (525), g d by Humphrey Davy (676), g g d by Own Brother to Bossy (925).

EMPEROR, (525), l. in 1855; br. J. Evans, s Own Brother to Bossy (925), d by Old Shrawardine (Crane).

ENSDON RESERVE, (526), l. in 1881; br. H. Lee & Son, s Royal Reserve (159), d by Young Caligula (1293).

ENSIGN, (527); br. C. Byrd, s Major (84), d by Canterbury Patentee 3rd (C. Byrd).

ERCALL P, (528), l. in 1879; br. J. Harding, s Pride of Montford (959), d a Mansell Ewe.

P. EVERALL'S No. 3, (529); br. P. Everall, s Cardinal 2nd (340)

EVANS' No. 12, 1867 (530), l. in 1866; br. J. Evans, s Nonpareil (908), d by Emperor (525).

EVANS' No. 7, 1869 (531), l. in 1868; br. J. Evans, s Nonpareil (908), d by Young Emperor (1305).

EVANS' No. 1, 1872 (532), l. in 1871; br. J. Evans, s Hardlines (647), d by Cardinal (53).

EVANS' No. 5, 1872 (533), l. in 1871; br. J. Evans, s Hardlines (647), d by Chieftain (384).

EVANS' No. 3, 1873 (534), l. in 1872 ; br. J. Evans, s Grand
Duke (620), d by Standard Bearer (80).

EVANS' RAM, (535) ; br. J. Evans.

EVANS' No. 12, 1875 (536), l. in 1874 ; br. J. Evans, s Claudius
(103), d by Nonpareil (908).

EVANS' RAM, (537), l. in 1874 ; br. J. Evans, s Grand Duke
(620), d by Standard Bearer (80).

EVERALL'S RAM, (538), l. in 1875 ; br. P. Everall, s Minton's
No. 1, 1874 (844), d by Son of Cardinal (1126).

EVERALL'S No. 16, 1882 (539), l. in 1881 ; br. P. Everall, s
Hercules (663), d by Prince Victor (158).

EXILE, (540), l. in 1866 ; br. R. H. Masfen, s Wanderer (R. H.
Masfen), d by The Rejected (R. H. Masfen).

FALSTAFF, (541), l. in 1880 ; br. J. Beach, s First Choice (561),
d by British Oak (290).

FARMERS' No. 1, 1878 (542), l. in 1877 ; br. J. E. Farmer, s Lord
Acton (731), d by Commonwealth (Masfen).

FARMER'S No. 2, 1878 (543), l. in 1877 ; br. J. E. Farmer, s
Beach's No. 3, 1873 (221), d by Sir Boughton (Farmer).

FARMER'S No. 15, 1878 (544), l. in 1877 ; br. J. E. Farmer, s
Beach's No. 3, 1873 (221), d by Sir Boughton (Farmer).

FARMER'S No. 10, 1880 (545), l in 1879 ; br. J. E. Farmer, s
Heart of Oak (659), d by Nightingale's No. 2.

FARMER'S No. 16, 1880 (546), l. in 1879 ; br. J. E. Farmer, s
Double X (501), d by Beach's No. 3, 1873 (221).

FARMER'S No. 36, 1881 (547), l. in 1881 ; br. J. E. Farmer, s
Double X (501), d by Beach's No. 3, 1873 (221).

FARMER'S FRIEND, (548), l. in 1876 ; br. Mrs. Wadlow, s a
Mansell Ram, d a Masfen Ewe.

FASCINATOR, (549) ; br. T. S. Minton, s Fenn's Ram, d a
Bromley Ewe.

FAT BACK, (550), l. in 1858 ; br. J. & E. Crane, s Tern (1176),
d a Crane Ewe.

FAT BACK, (551), l. in 1869 ; br. G. Allen, s Model Patentee.

FAT BACK, (552), l. in 1873 ; br. J. Coxon, s Stamina (1157),
d by Mansion 3rd (818).

FAVOURITE, (553), l. in 1868 ; br. R. H. Masfen, s Lord
Uffington (56), d by Brother to Gratitude (Masfen).

FAVOURITE (alias Cock of the Walk), (554), l. in 1874 ; br. J.
Coxon, s Combination, d a Keeling Ewe.

FENN'S No. 3, 1879 (555), l. in 1878 ; br. T. Fenn, s Beach's No.
17, 1876 (228), d by My Lord (Fenn).

FENN'S No. 4, 1879 (556), l. in 1878 ; br. T. Fenn, s Beach's 17, 1876 (228), d by My Lord (Fenn), g d by Lord Kenyon g g d by Novelty (41).

FENN'S No. 16, 1880 (557), l. in 1879 ; br. T. Fenn, s Ensdon Hero (104), d by Midlothian (842), g d by Marquis (820).

FENN'S No. 29, 1882 (558), l. in 1881 ; br. T. Fenn, s Clinker (397), d by the Buck (1178), g d by Lord Kenyon.

FIELD MARSHALL, (559), l. in 1872 ; br. C. R. Keeling, s Commander-in-Chief (416), d by Cardinal (53), g d by Havelock (27).

FIRMAMENT, (560), l. in 1880 ; br. T. Mansell, s North Star (913), d by Raby Duke (108).

FIRST CHOICE, (561), l. in 1878 ; br. T. Mansell, s County Member (452), d by Severn (1050).

FIRST CHOICE, (562), l. in 1882 ; br. J. Pickering, s Pride of Baschurch (953), d a Thomas Ewe.

FIRST FLIGHT, (563) ; br. R. Thomas, s A1 (176), d by Prince (963), g d by Co-Monument (423).

FIRST FRUITS, (564), l. in 1876 ; br. J. Coxon, s Colossus (411), d by Stamina (1157).

FLAG OF BRITAIN, (565), l. in 1881 ; br. E. Meredith, s British Yeoman (294), d by Long Back (727).

FLAG STAFF, (566), l. in 1880 ; br. R. Thomas, s Standard (1159), d by Grandeur (133), g d by Favourite (553).

FLOSS, (567), l. in 1880 ; br. G. Cooke, s Mrs. Smith's No. 7, 1877, d by a Mansell Ram.

FOREMAN, (568), l. in 1867 ; br. E. Thornton, s Earl Derby (T. Mansell), d by Patrician (C. P. Peters).

FORESTER, (570), l. in 1878 ; br. J. L. Naper, s Freebooter (Naper), d by Duke of Edinboro' (511).

FORESTER, (571), l. in 1881 ; br. A. P. Turner, s Farmer's No. 10, 1880 (545), d by a Stubbs Ram.

FORMATION, (572), l. in 1876 ; br. J. Coxon, s Champion (351), d by Stamina (1157), g d by Sheet Anchor (1060).

FORTON 1st, (573), l. in 1879 ; br. J. W. Minton, s His Lordship (669), d by Bedford Hero (111).

FORTON 2nd, (574), l. in 1881 ; br. J. Bowen-Jones, s Harding's No. 1, 1880.

FORTON HERO, (575), l. in 1876 ; br. J. W. Minton, s Bedford Hero (111), d by Montford (869).

FOSSE DUKE, (576) ; br. T. Horley, s Duke of Newcastle (39), d a Horley Ewe.

FOSTER'S LEEDS H. C. RAM, (577), l. in 1859; br. W. O. Foster.

FOSTER'S RAM, (578), l. in 1855; br. W. O. Foster.

FOUNDATION, (579), l. in 1876; br. J. Coxon, s Colossus (411), d by Confidence (101), g d by Mansion 3rd (818).

FOUNDATION, (580), l. in 1882; br. T. Nock, s Jumbo (Nock), d a H. Smith Ewe.

FRATERNITY (581), l. in 1880; br. J. E. Farmer.

FREEFORD, (582), l. in 1877; br. J. Coxon, s Magnet (788), d by Freeman (585), g d by Captivator (76).

FREEFORD, (583), l. in 1873; br. J. Coxon, s Captivator (76), d by Cavalier (Coxon), g d by Novelty (41).

FREEFORD DUKE, (584), l. in 1880; br. J. Coxon, s City Member (391), d by Marquis of Bath (822), g d by Confidence (101).

FREEMAN, (585), l. in 1876; br. J. Coxon, s Colossus (411), d by Captivator (76).

FRERE, (586), l. in 1881; br. G. Cooke, s Beaufort, d by Son of British Oak.

FRIAR TUCK, (587), l. in 1879; br. Crane & Tanner, s Columbus (128), d by Caligula (112), g d by Celebrity (6).

FRINGE, (588), l. in 1881; br. G. Cooke, s Beaufort, d by Son of British Oak.

FULL EYE, (589), l. in 1881; br. G. Cooke, s Son of British Oak, d by a Beach Ram.

GALLANT, (590), l. in 1879; br. J. W. Minton, s His Lordship (669), d by Bedford Hero (111).

GAMBETTA, (591), l. in 1881; br. T. J. Mansell, s Pride of Montford (959), d by Double B (500).

GAYTON PRINCE, (592), l. in 1863; br. S. Byrd, s Prince (J. Evans), d by Quality (S. Byrd).

G.C.B., (593), l. in 1880; br. T. S. Minton, s Marquis of Bath (822), d a Minton Ewe.

GENERAL, (594), l. in 1876; br. Viscount Falmouth, s Secundus (1049), d by Lord Paramount (772).

GENERAL, (595), l. in 1876; br. V. E. Nightingale, s Chesham's No. 9, 1875 (365), d by Son of Worcester (1149).

GENERAL, (596), l. in 1872; br. T. Mansell, s Rifleman (1007), d by Conservative (435).

GENERAL, (597), l. in 1880; br. J. Pulley, s Young Sultan (1328).

GENERAL, (598), l. in 1873 ; br. R. H. Masfen, s Cardigan (338), d by Young Battersea (Masfen), g d by Corinthian (Coxon), g g d by Brother to Gratitude (Masfen).

GENERALISSIMO, (599), l. in 1877 ; br. G. Graham, s Prince Victor (Graham), d a Keeling Ewe.

GENIUS, (600), l. in 1875 ; br. J. Evans, s Grand Duke (620), d by Broadgauge (89).

GEORGE IV, (601), l. in 1880 ; br. G. Graham, s Georgius (139), d a Graham Ewe.

GERMAN, (602), l. in 1874 ; br. W. German, s Confidence (101), d by Preserver (950).

GERMAN'S No. 6, 1875 (603), l. in 1874 ; br. W. German, s Consolation (438).

GERMAN'S No. 16, 1881 (604), l. in 1880 ; br. G. German, s Farmer's Friend (548), d by Ranger (992).

GIANT, (605), l. in 1879 ; br. M. Williams, s Hercules (663), d by Young Hardlines, g d by Exile (540).

GIGANTIC, (606), l. in 1879 ; br. C. Randell, s Colossal (410), d by Our Chief (924).

GIPSEY KING, (607), l. in 1873 ; br. Viscount Falmouth, s Nobleman (900), d a Mansell Ewe.

GLADIATOR, (608), l. in 1877 ; br. J. Coxon, s Champion (351), d by Utility (1253), g d by Novelty (41).

GLENDOWER, (609), l. in 1881 ; br. J. Evans, s Lord Coxcomb (743), d by May Duke (837), g d by Proud Salopian (87).

GOLD DUST, (610), l. in 1881 ; br. W. German, s Cockade (400), d by Marquis of Bath (822).

GOLIAH, (611) ; br. R. H. Masfen, s Commander-in-Chief (416), d by Corsair (57), g d by Corinthian (Coxon), g g d by Brother to Gratitude (Masfen).

GOLIATH, (612), l. in 1878 ; br. Viscount Falmouth, s Masterman (828), d a Beach Ewe.

GOLIAH 2nd, (613), l. in 1878 ; br. H. J. Sheldon, s Philistine (936), d by True Type (85).

GRAHAM'S No. 6, 1880 (614), l. in 1879 ; br. G. Graham, s Prince Imperial (966).

GRAHAM'S No. 7, 1880 (615) ; br. G. Graham, s The Star (Masfen).

GRAHAM'S No. 7, 1881 (616), l. in 1880 ; br. G. Graham, s Georgius (139), d a Graham Ewe.

GRAHAM'S No. 10, 1881 (617), l in 1880 ; br. G. Graham, s Graham's No. 12, 1880, d a Graham Ewe.

GRAND CHIEF, (618), l. in 1878; br. R. Thomas, s Grandeur (133), d by Calcot Chieftain (319), g d by Favourite (553).

GRAND CHIEF, (619), l. in 1881; br. J. Darling, s Royal Chief (1022), d by Grand Turk (629).

GRAND DUKE, (620), l. in 1870; br. J. Evans, s Cardinal (53), d by Volunteer (51), g d by Sir Samuel (1113), g g d by Emperor (525).

GRAND MASTER, (621), l. in 1879; br. R. Thomas, s Grand Chief (618), d by Co-Monument (423), g d by Favourite (553).

GRAND MASTER, (622), l. in 1873; br. J. Harding.

GRAND MASTER, (623), l. in 1875; br. J. Evans, s Grand Duke (620), d an Evans Ewe.

GRAND MASTER, (624), l. in 1879; br. Viscount Falmouth, s Masterman (828), d a Beach Ewe.

GRAND PRINCE, (625), l. in 1882; br. R. Thomas, s Grand Master (621), d by Prince (963), g d by Son of Conservative (1133).

GRAND SIEGNEUR, (626), l. in 1879; br. J. Darling, s Grand Turk (629), d by Beaudesert (J. Darling).

GRANDSON OF OXFORD HERO, (627), l. in 1873; br. Lord Chesham, s Son of Oxford Hero, (Chesham) d by Radford.

GRANDSON OF NOBLEMAN, (628), l. in 1868; br. J. Coxon, s Commander (415), d by Nobleman (37).

GRAND TURK, (629), l. in 1877; br. G. Graham, s Prince Imperial (966), d a Firmstone Ewe.

GRAND VIZIER, (630), l. in 1874; br. J. Evans, s Grand Duke (620), d by Broadgauge (89), g d by Humphrey Davy (676).

GRAPHIC, (631), l. in 1877; br. G. Graham, s Prince Imperial (966), d by Patrician (120)

GREEN'S No. 22, (632), l. in 1878; br. J. B. & G. H. Green, s Green's No. 1 (Green), d by Nightingale's No. 1, 1873 (894).

GREEN'S No. 21, 1879 (633), l. in 1878; br. J. B. & G. H. Green, s Green's No. 2 (Green), d a Marlow Ewe.

GREEN'S No. 18, 1881 (634), l. in 1880; br. J. B. & G. H. Green, s Son of Ensdon Hero, d by Nightingale's No. 1, 1873 (894).

GREEN'S No. 12, 1880 (635), l. in 1879; br. J. B. & G. H. Green. s Acton, d a Marlow Ewe.

GREEN'S No. 6, 1881 (636), l. in 1881; br. J. B. & G. H. Green, s Green's No. 7 (Green), d by Pilot.

GREEN'S No. 12, 1881 (637), l. in 1881 ; br. J. B. & G. H. Green, s Farmer's No. 1, 1878 (542), d a Marlow Ewe.

GREEN'S No. 10, 1832 (638), l. in 1881 ; br. J. B. & G. H. Green, s Farmer's No. 1, 1878 (542), d by Green's No. 22 (632).

GREEN'S No. 12, 1882 (639), l. in 1881 ; br. J. B. & G. H. Green, s Fenn's No. 16, 1880 (557), d a Marlow Ewe.

GREEN'S No. 18, 1882 (640), l. in 1881 ; br. J. B. & G. H. Green, s Fenn's No. 16, 1880 (557), d by Nightingale's No. 1, 1873 (894)

GRINDLE 2nd, (641), l. in 1880 ; br. W. Yates, s Grindle, d by Lord Dartmouth's No. 38, g d by Mansell's No. 4, g g d by Plymouth.

GUARDIAN, (642) ; br. C. Byrd, s Canterbury 2nd (328).

HAND'S RAM, (643), l. in 1860 ; br. J. Hand, d a Wigley Ewe.

HAND'S RAM, (644) ; br. W. O. Foster.

HAPPY-GO-LUCKY, (645) ; br. Viscount Falmouth, s Luck's All (782), d by Lord Paramount (772).

HARDING'S No. 2, 1881 (646), l. in 1880 ; br. J. Harding, s Caligula 2nd (322), d by Mansell's No. 16, 1876 (802).

HARDLINES, (647), l. in 1869 ; br. J. Evans, s Nonpareil (908), d by Young Emperor (1305), g d by Sir Samuel (1113), g g d by Humphrey Davy (676).

HARD TIMES, (648), l. in 1879 ; br. J. Harding, s Mansell's No. 16, 1876 (802), d by the Buck (1178).

HARDY, (649), l. in 1878 ; br. R. Bromley, s Lord Harlech (135), d by Son of Hardlines.

HARDY, (650), l. in 1879 ; br. Mrs. H. Smith, s Multum in Parvo (885), d by Lord Hordsley.

HARLESCOTT, (651), l. in 1876 ; br. R. Barber, s an Evans Ram.

HARLESCOTT, (652), l. in 1878 ; br. R. Barber, s Crane's No. 1, 1877, d by Pride of Harlescott, g d by Lord Harlescott.

HARLESCOTT 2nd, (653), l. in 1881 ; br. M. Williams, s Harlescott (651), d by Bristol Reserve (144).

HARLESCOTT 2nd, (654), l. in 1881 ; br. M. Williams, s Harlescott (651), d an Evans Ewe.

HAROLD 2nd, (655), l. in 1880 ; br. R. Bromley, s Harold, (R. Bromley), d by Hardlines (647).

HARRY, (656), l. in 1871 ; br. Viscount Falmouth, s Tamworth (1172), d a Bradburne Ewe.

HATTON'S BRITISH OAK, (657), l. in 1877, br. Mrs. S. Beach, s British Oak, (290), d by Briton (296), g d by Young Gayton (1307).

HAWKINS' No. 6, 1880 (658), l. in 1879, (658); br. Crane & Tanner, s Columbus (128).

HEART OF OAK, (659), l. in 1878; br. J. E. Farmer, s Beach's No. 17. d a Smith Ewe.

HEART OF OAK, (660), l. in 1880; br. Viscount Falmouth, s Hatton's British Oak (657), d a Chesham Ewe.

HENRY SMITH, (661), l. in 1859; br. H. Smith, d a Gloucester 1st Prize Ewe.

HERCULES, (662), l. in 1872; br. R. H. Masfen, s True Type (85), d by Cardinal (53), g d by Compton Buck (A. H. Minor), g g d by Lady Jane Grey's Son.

HERCULES, (663), l. in 1878; br. M. Williams, Junr., s Bristol Reserve (144), d by Co-Monument (422), g d by Exile (540).

HEREFORD, (664), l. in 1879; br. J. Pulley, s Colossal (Pulley), d by Sultan (1167).

HERMIT, (665), l. in 1868; br. Mr. Johnson, s Son of Conservative, d by Cardinal (53).

HERMIT, (666), l. in 1879; br. C. Byrd, s The Hero (1188), d by Hampton Hero (84).

HERO, (667), l. in 1868; br. Mrs. S. Beach, s Cardinal (53).

HERO, (668), l. in 1881; br. T. S. Minton, s Montford Hero (160), d by His Lordship (669).

HIS LORDSHIP, (669), l. in 1877; br. T. Mansell, s Raby Duke (108), d by Landseer (700).

HIS LORDSHIP 2nd, (670), l. in 1878; br. T. Mansell, s Raby Duke (108), d by Landseer (700).

HOPEFUL, (671), l. in 1877; br. T. Ryland, s German's No. 18, 1876, d a Crane Ewe.

HORTON'S BIG RAM, (672), l. in 1863; br. T. Horton, s Lord Salop (23), d by Lord Suffolk (17).

HORLEY'S No. 29, 1870 (673), l. in 1869; br. T. Horley, Junr., s Evans' No. 10, 1867, d by Duke of Newcastle (39), g d by Masfen's No. 6.

HORTON'S No. 16, 1861 (674), l. in 1860; br. T. Horton, s Young Patentee (1320).

HUBERT, (675), l. in 1882; br. Sir C. F. Smythe, Bart., s a Mansell Ram, d a Hudson Ewe.

HUMPHREY DAVY, (676), l. in 1857; br. S. Meire, s Magnum Bonum (Meire), d by Bossy (Meire)

IMPROVER, (677), l. in 1875; br. Mrs. Smith, s Young Creole.

IMPROVER, (678), l. in 1881 ; br. J. Darling, s Dudmaston Hero (165), d by Grand Turk (629).

INSTONE'S RAM, (679), l. in 1881 ; br. E. Instone, s Crane & Tanner's No. 16, 1880, d by Shropshire Ram (134).

IPSLEY'S PRIDE, (680), l. in 1882 ; br. E M. Coleman, s Mansell's No. 12, 1881 (811), d a Minton Ewe.

IRON DUKE, (681) ; br. R. H. Masfen, s Pride of Pendeford (Masfen).

IRONMASTER, (682), l. in 1878 ; br. Mrs. S. Beach, s British Oak (290), d by Cardigan (338), g d by Young Gayton (1307).

JACK TAR, (683), l. in 1876 ; br. J. Evans, s British Oak (290), d by Union Jack (1252) ; g d by Emperor (525).

J. BOWEN-JONES' No. 6, 1881 (684), l. in 1880 ; br. J. Bowen-Jones, s Dudmaston 21, 1878 (507), d by Crane's No. 6, 1876 (470).

J. BOWEN-JONES' RAM (685), l. in 1877 ; br. J. Bowen-Jones, s Calcot (318), d by Lord Warden (777).

JEMMY, (686), l. in 1871 ; br. Viscount Falmouth, s Yates' No. 6, 1870 (1285), d a Yates Ewe.

J. HARDING'S No. 10, 1831 (687), l. in 1880 ; br. J. Harding, s Caligula 2nd (323), d by Mansell's No. 16, 1876 (802).

JOCK, (688), l. in 1882 ; br. T. Nock, s Jumbo (T. Nock), d by Sandboy (1044).

JUCKES' RAM, (689), l in 1853 ; br. Mr. Juckes, Tern.

JULIUS, (690), l. in 1867 ; br. T. Horley, s Duke of Newcastle (39), d by Duke of Kent (13).

JULIUS, (691), l. in 1876 ; br. T. Nock, s Mansell's No. 9, d by Creole (Coxon).

JULIUS CÆSAR, (692), l. in 1876 ; br. Mr. Picken, s Cæsar (Evans), d by a Beach Ram.

K.C.B., (693), l. in 1880 ; br. T. S. Minton, s Marquis of Bath (822), d by Lord Clifden (44).

KENWICK, (694), l. in 1879 ; br. R. Thomas, s Grand Chief (618), d by Benthall Chieftain (246), g d by Castle Warden (342).

KENYON, (695), l. in 1867 ; br. The Hon. E. Kenyon, s Model, d by Young Sir Harry.

KING HARRY, (696), l. in 1875 ; br. Viscount Falmouth, s Sheriff (1073), d by Harry (656).

KING OF THE HEATH, (697), l. in 1880 ; br. W. Ward, s M. Williams's No. 1, 1879 (1273), d by Bristol Reserve (144), g d by Claude Duval (394), g g d by Claudius (103).

KINGSCRAFT, (698), l. in 1869; br. T. Fenn, s Novelty (41), d a Fenn Ewe.

KING WILLIAM, (699), l. in 1881; br. M. Williams, s Pride of Bishton (954), d by Artist (194).

LANDSEER, (700), l. in 1872; br. E. Crane, s Cambrian (324), d by Nobleman (37).

LATIMER, (701), l. in 1871; br. The Rt. Hon. Lord Chesham.

LATIMER, (702), l. in 1881; br. Lord Chesham, s Grandson of Mansell's No. 6, d by Primate (Evans).

LATIMER WONDER, (703), l. in 1881; br. Lord Chesham, s Dudmaston (506), d by Royal Aston (124).

LAVENDER, (704), l. in 1881; br. T. Nock, s Cynic 2nd, d by Chesham's No. 5, 1873.

LEE'S No. 1, 1880 (705); br. H. Lee & Son.

LEINSTER, (706), l. in 1875; br. R. H. Masfen, s Patrick (Hamilton), d by Grindle (Keeling), g d by Corinthian (Coxon), g g d by Mainstay (Masfen).

LEO, (707), l. in 1878; br. G. W. Wheeler, s Mrs. Smith's No. 12, 1875 (1118), d by Lord Chesham's No. 19 (364), g d by Posenhall (942).

LEVIATHAN, (708), l. in 1872; br. E. Lythall, s Warrior (E. Lythall).

LEVIATHAN, (709), l. in 1873; br. T. Mansell, s Calcot (317), d by Conservative (435).

LEVIATHAN, (710), l. in 1874; br. R. H. Masfen, s Consequence (Bradburne), d by Magdala (Masfen), g d by Corinthian (Coxon), g g d by Mainstay (Masfen).

LEVIATHAN 2nd, (711); br. H. Matthews, s Leviathan (72), d a Matthews Ewe.

LEVIATHAN 3rd, (712), l. in 1874; br. H. Matthews, s Leviathan 2nd (711), d a Matthews Ewe.

LEVIATHAN 4th, (713), l. in 1879; br. R. Meredith, s Leviathan 3rd, (712), d by The Tory (1208).

LICHFIELD, (714), l in 1873; br, John Coxon. s Stamina 1157), d by Mansion 3rd (818), g d by Novelty (41.

LITTLE CHESTER, (715), l. in 1861; br. Viscount Falmouth, s Chester (381), d a R.A.S.E. Ewe.

LITTLE JOHN, (716), l. in 1881; br. The Rt. Hon. Lord Chesham, s Son of Colossus (Chesham), d by Royal Taunton (115), g d by Hon. Kenyon's Ram.

LITTLE JOHN, (717), l. in 1882; br. G. C. Price, s Robin Hood 2nd (1015), d by Clansman (G. C. Price), g d by Milton (Chesham).

LITTLE LORD, (718) ; br. T. Mansell, s Conservative (435), d by Earl Plymouth (48).

LITTLE MINTON, (719) ; br. T. S. Minton, s Marquis of Bath (822), d by Bedford Hero (111), g d a Minton Ewe.

LITTLE NIGHTINGALE, (720), l. in 1879 ; br. J. B. & G. H. Green, s Green's No. 2 (Green), d by Mansell's No. 21, 1874 (799).

LITTYWOOD PATENTEE, (721) ; s Canterbury 2nd (328), d by Young Excelsior.

LOADSTONE, (722), l. in 1881 ; br. R. Loder, M.P., s Royal Victor (1030), d a Chesham Ewe.

LODER'S No. 1, 1881 (723), l. in 1880 ; br. R. Loder, M.P., s a Chesham Ram, d a Chesham Ewe.

LODER'S No. 11, 1882 (724), l. in 1881 ; br. R. Loder, M.P , s Royal Victor (103C), d a Chesham Ewe.

LODER'S RAM, (725), l. in 1878 ; br. R. Loder, M.P., s a Chesham Ram.

LODER, (726), l. in 1881 ; br. R. Loder, M.P., s Victor (150), d a Chesham Ewe.

LONG BACK, (727), l. in 1871 ; br. G. Allen, s Fat Back (551), d by Nonpareil (908).

LONGBOW, (728), l. in 1874 ; br. T. Mansell, s Longitude (729), d by Mansion 2nd (55).

LONGITUDE, (729), l. in 1870 ; br. T. Mansell, s Pattern (933), d by Mansion 2nd (55).

LONG SHEEP, (730), br. Viscount Falmouth, s a Mansell Ram, d a Falmouth Ewe.

LORD ACTON, (731), l. in 1873 ; br. C. Wadlow, s an Instone Ram, d a Crane Ewe.

LORD ALCESTER, (732), l. in 1882 ; br. C. Randell, s Collingwood (407), d by Julius Cæsar (692).

LORD ALFRED, (733), l. in 1880 ; br. J. Darling, s Monarch (864), d a Darling Ewe.

LORD APLEY, (734) l. in 1874 ; br. W. O. Foster, s Lord Hull (100), d by Conservative (435).

LORD ASTLEY, (735), l. in 1860 ; br. A. H. Minor, s Earl of Liverpool (H. Smith), d by Emperor (525).

LORD BEACONSFIELD, (736), l. in 1875, br. T. J. Mansell, s Double B. (500), d by Little Lord (718).

LORD BEACONSFIELD, (737), l. in 1877 ; br. J. H. Bradburne, s Lord Aston (123), d by Crosswood Hero 2nd.

LORD BLAYDON (738), 1 in 1877 ; br. H. J. Sheldon, s Taunton Reserve (118), d a Sheldon Ewe.

LORD BRAILES, (739), l. in 1879 ; br. H. J. Sheldon, s Model (862).

LORD CHANCELLOR, (740), l. in 1881 ; br. J. Evans, s Lord Coxcomb (743), d by Grand Duke (620), g d by Claudius (103).

LORD CLAUD, l. in 1875 (741) ; br. E. Crane, s Claudius, d by Lord Uffington.

LORD CLIVE, (742), l. in 1879 ; br. J. W. Minton, s His Lordship (669), d by Calcot (317).

LORD COXCOMB, (743), l. in 1879 ; br. J. W. Minton, s Son of Coxcomb (Crane), d by Bedford Hero (111).

LORD DE LISLE, (744), l. in 1881 ; br. C. H. Clark, s Lord Leicester (755), d by Coxon's No. 14, 1878 (461), or His Lordship 2nd (670).

LORD EXETER, (745), l. in 1877 ; br. T. J. Mansell, s Beach's No. 17, 1876 (228), d by Son of Lord Clifden (1141).

LORD FALMOUTH, (746), l. in 1880 ; br. J. Beach, s First Choice (561), d by British Oak (290), g d by Lord Stafford (775).

LORD FLASH, (747), l. in 1857 ; br. J. & E. Crane, s Tern (1176), d a Crane Ewe.

LORD FOPPINGTON, (748), l. in 1881 ; br. J. Evans, s Lord Coxcomb (743), d by May Duke (837), gd by Broadguage (89).

LORD GREY, (749) ; br. C. Byrd, s Lord Weston (778), d by Constitution (S. Byrd).

LORD HATTON, (750), l. in 1880 ; br. J. Beach, s First Choice (561), d by Cardigan (338).

LORD HAUGHTON, (751), l. in 1870 ; br. Mrs. Wadlow, s Mansion 2nd (55), d Sister to Conservative (435).

LORD KINETON, (752), l. in 1877 ; br. Lord Willoughby de Broke, s Son of Lord Hull, d a Willoughby de Broke Ewe.

LORD KNIGHTLEY, (753) ; br. T. & T. J. Mansell, s Conservative (435), d by Earl Plymouth (48).

LORD LATIMER, (754) ; br. Lord Chesham, s Lord Kingston (99).

LORD LEICESTER, (755), l. in 1879 ; br. Mrs. Barrs, s The Czar (1182), d by Major (791).

LORD LINCOLN, (756) ; br. J. Coxon, s Duke of Newcastle (39), d by Corrector.

LORDLY, (757), l. in 1881 ; br. T. J. Mansell, s Forton (573), d by Warwick (1268).

58

LORD MAYOR, (758); br. Viscount Falmouth, s Sheriff (1073).

LORD MILCOMBE, (759), l. in 1881; br. E. Johnson, M.P., s Sultan of Farringdon (Pulley), d by a Chesham Ram, g d a H. Smith Ewe, g g d a Hawkins Ewe.

LORD MILFORD, (760), l. in 1868; br. T. Mansell.

LORD MINTON, (761), l. in 1880; br. T. S. Minton, s Marquis of Bath (822), d by Son of Bedford Hero (1123).

LORD MONA, (762), l. in 1876; br. G. German, s Conserver (437), d by Consolation.

LORD MONTFORD, (763), l. in 1880; br. T. S. Minton, s Marquis of Bath (822), d a Minton Ewe.

LORD NAPIER, (764); br. C. Byrd, s Lord Weston (778), d a Byrd Ewe.

LORD OAKENSHAW, (765), l. in 1881; br. E. M. Coleman, s Cyprus (486), d a Minton Ewe.

LORD ODSTONE, (766), l. in 1876; br. Fenn & Harding, s Ensdon Hero (104), d by Midlothian (842), g d by Marquis (820), g g d by Lord Kenyon.

LORD OF THE HAREM, (767), l. in 1881; br. Crane & Tanner, s Dudmaston (504), d by Claudius (108).

LORD OF THE HEATH, (768), l. in 1880, br. Wm. Ward, s Son of Prince (1145), d by Prince (963), g d by Claudius (103); g g d by Caractacus (335).

LORD OF THE ISLES, (769), l. in 1860; br. T. Horton, s Lord Suffolk (17), d by Shropshire, (Horton).

LORD OF THE ISLES, (770), l. in 1876; br. J. Harwood, s Waverley, d by Caradoc (336).

LORD OF THE MANOR, (771), l. in 1878; br. G. German, s Marquis of Bath (822), d by Captivator (76).

LORD PARAMOUNT, (772); br. C. Byrd, s Knight Templar, d by Lord Clifden (44).

LORD RADFORD, (773), l. in 1873; br. S. C. Pilgrim, s Commandant (S. Pilgrim), d by Primate (S. Pilgrim).

LORD RUGBY, (774), l. in 1880; br. E. M. Coleman, s Lord of the Isles (770), d a Minton Ewe.

LORD STAFFORD, (775), l. in 1871; br. Mrs. S. Beach, s Duke of Manchester (70), d by Gayton Prince (592).

LORD STANLEY, (776); br. E. Thornton, s Earl Derby (T. Mansell), d by Lord Clifden (44), g d by Pride of Pitchford (961), g g d by Canterbury.

LORD WARDEN, (777), l. in 1862; br. P. W. Bowen, s Comet (Crane), d by Chester Billy (7).

LORD WESTON, (778) ; br. Mr. Byrd, s Constitution (S. Byrd).

LOTHAIR, (779), l. in 1876 ; br. Lord Chesham, s Marquis of Bute (91), d by Oxford Hero (77).

LOWER EATON, (780), l. in 1881 ; br. E. Johnstone, M.P., s Sultan of Farringdon (Pulley), d by Cannock Chief (326), g d by Czar (Pulley).

L.S.D., (781), l. in 1877 ; br. Mrs. S. Beach, s Sterling (Beach), d by Monarch (863), g d by Cardinal (53).

LUCK'S ALL, (782) ; br. R. H. Masfen, s Coxswain (Masfen), d a Keeling Ewe.

LUSTY, (783), l. in 1881 ; br. J. E. Farmer, s Carlisle (166), d by Beach's No. 3, 1873 (221).

M.A., (784), l. in 1877 ; br. Mrs. Beach, s Masterman (828), d by British Oak (290), g d by Duke of Manchester (70).

MACCARONI, (785) ; br. T. Mansell, s Mansell's No. 6, 1859 (797), d a Mansell Ewe.

MACKINTOSH, (786), l. in 1880 ; br. T. Mansell, s County Member (452), d by Artist (194).

MAGNET, (787), l. in 1872 ; br. Mr. Walker, s Claudius (103), d by Perfection.

MAGNET, (788), l. in 1875 ; br. T. Mansell, s Pattern (933), d by Longitude (729).

MAGNUM, (789), l. in 1881 ; br. R. M. Knowles, s Multum-in-Parvo (885), d by Lord Mayor (146).

MAGNUM BONUM, (790), l. in 1880 ; br. T. J. Mansell, s Multum-in-Parvo (885), d a Crane Ewe.

MAJOR, (791), l. in 1872 ; br. Mr. Walker, s Claudius (103), d by Young Duke.

MAJOR, (792), l. in 1875 ; br. C. R. Keeling, s Captain (R. H. Masfen), d by Commonwealth (R. H. Masfen).

MAJOR, (793), l. in 1880 ; br. T. Mansell, s Multum-in-Parvo (885), d by County Member (452).

MANAGER, (794) ; br. T. Mansell, s Conservative (435), d by Big Neck (T. Mansell).

MANAGER, (795), l. in 1875 ; br. T. J. Mansell, s son of Conservative (1134), d by Calcot (317).

MANSELL, (796), l. in 1881 ; br. T. J. Mansell, s Caligula 2nd, (322), d by County Member (452).

MANSELL'S No. 6, 1859 (797) ; s Tame Deer (Mansell), d a Dyott Ewe

MANSELL'S No. 5, 1860 (1226), l. in 1859 ; br. T. Mansell, s Short-legged Patentee (1076), d a Mansell Ewe.

MANSELL'S No. 6, 1868 (798), l. in 1867 ; br. T. Mansell, s
Conservative (435), d by Lord Clifden (44).

MANSELL'S No. 21, 1874 (799), l in 1873 ; br. T. & T. J.
Mansell, s Severn (1050), d by Conservative (435).

MANSELL'S No. 7, 1875 (800), l. in 1874 ; br. T. Mansell, s
Landseer (700), d by Pattern (933).

MANSELL'S No 8, 1876 (801), l. in 1875 ; br. T. J. Mansell, s
Type (T. J. Mansell), d by Calcot (317).

MANSELL'S No. 16, 1876 (802), l. in 1875 ; br. T. Mansell, s
General (596), d by Marquis (820).

MANSELL'S No. 1, 1877 (803), l. in 1876 ; br. T. J. Mansell, s
Double B. (500), d by Calcot (317).

MANSELL'S No. 4, 1877 (1224), l. in 1876 ; br. T. J. Mansell, s
Truestock (1243), d by Severn (1050).

MANSELL'S No. 10, 1877 (804), l. in 1876 ; br. T. J. Mansell, s
May Duke (837), d by Little Lord (718).

MANSELL'S No. 28, 1877 (805), l. in 1876 ; br. T. J. Mansell, s
May Duke (837), d by Calcot (317).

MANSELL'S No. 20, 1878 (806), l. in 1877 ; br. T. J. Mansell, s
May Duke (837), d by True Stock (1243), g d by Earl of
Plymouth (48).

MANSELL'S No. 22, 1878 (807), l. in 1877 ; br. W. F. Firmstone,
s Second to None (Chesham).

MANSELL'S No. 2, (808) ; br. T. Mansell, s Marquis (820), d by
Conservative (435).

MANSELL'S No. 15, 1879 (1225), l. in 1878 ; br. T. J. Mansell,
s Double B. (500), d by True Stock (1243), g d by Calcot
(317).

MANSELL'S No. 7, 1880 (809), l. in 1879 ; br. T. J. Mansell, s
Pride of Montford (959), d by Double B (500) g d by True
Stock (1243).

MANSELL'S No. 11, 1881 (810), l. in 1880 ; br. T. Mansell, s
Multum-in-Parvo (885), d by Calcot (317).

MANSELL'S No. 12, 1881 (811), l in 1880 ; br. T. J. Mansell, s
Warwick (1268), d a Crane & Tanner Ewe.

MANSELL'S No. 3, 1882 (812), l. in 1880 ; br. T. J. Mansell, s
Milton (843), d by Raby Duke (108).

MANSELL'S No. 7, 1882 (813), l. in 1881 ; br. T. J. Mansell, s
Colston (413), d by May Duke (837).

MANSELL'S No. 8, 1882 (814). l. in 1881 ; br. T. J. Mansell, s
Pride of Montford (959), d by His Lordship (669), g d by
Son of Conservative (1134).

MANSELL'S No. 15, 1882 (815), l. in 1881 ; br. T. J. Mansell, s Milton (843), d by True Stock (1243).

MANSELL'S NEWCASTLE RAM, (816), l. in 1863 ; br. T. Mansell, s Short Legged Patentee (1076).

MANSELL'S RAM, (817), l. in 1872 ; br. T. Mansell, s Little Lord (718), d by Conservative (435).

MANSION 3rd, (818), l. in 1867 ; br. T. Mansell, s Conservative (435), d by Young Buckskin (1291).

MARK ANTHONY, (819), l. in 1878 ; br. G. W. Wheeler, s Brutus (306), d by Chesham's No. 19 (364).

MARQUIS (820), l. in 1865 ; br. T. Mansell, s Lord Clifden (44), d by Mansell's No. 6, 1859 (797).

MARQUIS, (821) ; br. Lord Willoughby de Broke, s The Young Duke (1211).

MARQUIS OF BATH, (822), l. in 1876 ; br. T. Mansell, s Artist (194), d by Longitude (729), g d by Conservative (435).

MARQUIS OF MONTFORD, (823), l. in 1880 ; br. T. S. Minton, s Marquis of Bath (822), d by Son of Conservative (1134).

MARSHALL, (824), l. in 1877 ; br. T. Fenn, s Mansell's No. 16, 1876 (802), d by Ensdon Hero (104), g d by Marquis (820).

MARSHALL FREEFORD, (825), l. in 1871 ; br. J. Coxon, s Warrior, d by Duke of Newcastle (39).

MASTER BEACH, (826), l. in 1877 ; br. Mrs. S. Beach, s Masterman (828), d by Duke of Manchester (70), g d by Cardinal (53).

MASTER DICK, (827), l. in 1880 ; br. Crane & Tanner, s Yardley (1283), d by Caligula (112), g d by Cato (116), g g d by Celebrity (6).

MASTERMAN, (828), l. in 1874 ; br. R. H. Masfen, s His Majesty (T. Mansell), d by Commander-in-Chief (416), g d Keeling's No. 19, g g d The Rejected.

MASTERMAN, (829), l. in 1877 ; br. E. Meredith, s The Tory (1208), d by Long Back (727).

MASTERMAN, (830), l. in 1879 ; br. G. German, s Magnet (788), d by Confidence (101).

MASTERPIECE, (831), l. in 1878 ; br. Viscount Falmouth, s Masterman (828), d a Beach Ewe.

MASTERPIECE, (832), l. in 1876 ; br. J. Coxon, s Magnet (788), d by Freeman (585).

MASTER MAY, (833), l. in 1879 ; br. J. Evans, s May Duke (837), d by Union Jack (1252), g d by Cardinal (53).

MATCHLESS, (834), l. in 1874 ; br. Lord Chesham, s Marquis of Bute (91) d by a Nock Ram.

MATTHEWS' No. 1, 1876 (835), l. in 1875 ; br. H. Matthews, s Montford Hero (Matthews), d by Young Leviathan (Matthews).

MAXIMUS, (836), l. in 1881 ; br. T. S. Minton, s Milton (843), d by Marquis of Bath (822).

MAY DUKE, (837), l. in 1874 ; br. J. Evans, s Grand Duke (620), d by Nonpareil (908) ; g d by Young Emperor (1305), g g d by Humphrey Davy (676).

MAY-FLY, (838), l. in 1879 ; br. J. H. Bradburne, s. Coxcomb, (Bradburne), d a Bradburne Ewe.

MEMBER FOR HEREFORD, (839), l. in 1881 ; br. E. Johnston, M.P., s Sultan of Farringdon (Pulley), d by Cannock Chief (326), g d by Czar (Pulley).

MERRY MONARCH, (840), l. in 1880 ; br. J. Evans, s May Duke (837), d by True Light (1242), g d by Royalty (1029).

MIDLAND, (841), l. in 1875 ; br. T. J. Mansell, s Son of Severn, d by Pattern (933).

MIDLOTHIAN, (842), l. in 1871 ; br. R. H. Masfen, s Rob Roy (1018), d by Milton (Masfen).

MILTON, (843), l. in 1878 ; br. T. J. Mansell, s County Member (452), d by Severn (1050). g d by Rifleman (1007).

MINTON'S No. 1, 1874 (844), l. in 1873 ; br. J. W. Minton, s Foreman (568), d a Minton Ewe.

MINTON'S No. 6, 1876 (845), l. in 1875 ; br. J. W. Minton, s Bedford Hero (111), d by Son of Lord Clifden (1141).

MINTON'S No. 12, 1877 (846), l. in 1876 ; br. J. W. Minton, s Bedford Hero (111), d by Little Lord (718).

MINTON'S No. 1, 1878 (847), l. in 1877 ; br. T. S. Mintford.

MINTON'S No. 14, 1880 (848), l. in 1880 ; br. T. S. Minton, s Marquis of Bath (822), d by Bedford Hero (111).

MINTON'S No. 20, 1880 (849), l. in 1879 ; br. J. W. Minton, s His Lordship (669), d by Bedford Hero (111).

MINTON'S No. 6, 1881 (850), l. in 1880 ; br. T. S. Minton, s Son of Conservative (1134), d a Minton Ewe.

MINTON'S No. 9, 1881 (851), l. in 1880 ; br. T. S. Minton, Marquis of Bath (822), d by Son of Conservative (1134).

MINTON'S No. 14, 1881 (852), l. in 1880 ; br. T. S. Minton, s Marquis of Bath (822), d by Bedford Hero (111).

MINTON'S No. 20, 1881 (853), l. in 1880 ; br. T. S. Minton, s Marquis of Bath (822).

MINTON'S No. 24, 1881 (854), l. in 1880 ; br. T. S. Minton, s Marquis of Bath (822), d a Minton Ewe.

MINTON'S No. 25, 1881 (855), l. in 1880 ; br. T. S. Minton, s Marquis of Bath (822), d by Severus (T. J. Mansell).

MINTON'S No. 10, 1882 (856), l. in 1881 ; br. T. S. Minton, s Montford Hero (160), d by Bedford Hero (111).

MINTON'S No. 13, 1882 (857), l. in 1881 ; br. T. S. Minton, s Montford Hero (160), d a Minton Ewe.

MINTON'S PRIDE, (858), l. in 1880 ; br. T. S. Minton, s Marquis of Bath (822), d a Minton Ewe.

MINTON'S PRIZE, (859), l. in 1877 ; br. J. W. Minton, s Son of Bedf rd Hero (1123), d a Forton Ewe.

MINTON'S RAM, (860) ; br. J. W. Minton.

MITRE, (861), l. in 1878 ; br. J. Darling, s Martyr, (Darling), d by Bradburne's No. 3, 1875.

MODEL, (862), l. in 1877 ; br. H. J. Sheldon, s Taunton Reserve (118), d a Sheldon Ewe.

MONARCH, (863), l. in 1874 ; br. Mrs. S. Beach, s Reflection (998), d by Cardinal (53).

MONARCH, (864), l. in 1876 ; br. Mrs. S. Beach, s Masterman (828), d by Duke of Manchester (70), g d by Byrd's No. 24, 1865.

MONARCH, (865), l. in 1877 ; br. J. Coxon, s Leviathan (710), d by Champion (351), g d by Sheet Anchor (1060).

MONARCH, (866) ; s Quality (S. Byrd).

MONARCH, (867), l. in 1881 ; br. T. S. Minton, s Son of Pride of Montford, alias Ercall P. (528). d a Minton Ewe

MONITOR, (868), l in 1881 ; br. T. S. Minton, s Montford Hero (160), d by Son of Bedford Hero (1123).

MONTFORD, (869), l. in 1868 ; br. H. Matthews.

MONTFORD (870), l. in 1877 ; br. J. W. & T. S. Minton.

MONTFORD, (871), l. in 1881 ; br. T. S. Minton, s Ercall P. alias Son of Pride of Montford (528), d by Son of Bedford Hero (1123).

MONTFORD No. 8, 1882 (872), l. in 1881 ; br. T. S. Minton, s Milton (843), d by Bedford Hero (111).

MONMOUTH, (873), l. in 1875 ; br. T. Nock, s Mr. Evans's No. 32, 1874, d by Blood Royal (83).

MOUNTAIN, (874), l. in 1881 ; br. W. German, s Masterman (828), d by Colossus (411).

MOUNTAIN 2nd, (875), br. P. W. Bowen, s Mountain (874), d by Patentee the Prime (28).

MOUNTAIN 3rd, (876), l. in 1869 ; br. P. W. Bowen, s Mountain 2nd (875), d a Pryce W. Bowen Ewe.

MOUNTAINEER, (877), l. in 1879; br. J. Evans, s May Duke (837), d by Cavalier (343), g d by Proud Salopian (87).

MOUNTAIN HERO, (878), l. in 1871 ; br. E. Meredith.

M.P., (879), l. in 1878; br. T. Mansell, s County Member (452), d by Pattern (933).

M.P., (880), l. in 1881.; br. T. S. Minton, s Milton (843), d by Double B. (500).

M.P., (881), l. in 1881 ; br. Col. Dyott.

MULTUM, (882), l. in 1880; br. T. Mansell, s Multum-in-Parvo (885), d by Landseer (700).

MULTUM, (883), l. in 1881 ; br. R. M. Knowles, s Multum-in-Parvo (885), d a Coxon Ewe.

MULTUM-IN-PARVO, (884), l. in 1868 ; br. W. German, s Lord Lincoln (German), d by Elford.

MULTUM IN-PARVO, (885), l. in 1877 ; br. T. J. Mansell, s Double B. (500), d by Little Lord (718), g d by Rifleman (1007).

MUTINEER, (886), l. in 1880; br. J. Evans, s May Duke (837), d by Union Jack (1252), g d by Proud Salopian (87).

MYSTERY, (887), l. in 1880 ; br. Lord Chesham, s M.A. (784), d a Chesham Ewe.

NEGRO (888), l. in 1864; br. C. W. Hamilton, s Jack, d a Hamilton Ewe.

NELSON, (889), l. in 1877 ; br. Mrs. S. Beach, s British Oak (290), g d by Cardigan (338), g d by Gayton Prince (592).

NEPTUNE, (890), l. in 1871 br. J. Coxon, s Pilot (T. Mansell) d by Sheet Anchor (1060).

NEWARK, (891), l. in 1876 ; br. Lord Chesham.

NEWPORT JOHN, (892), l. in 1878 ; br. P. Everall.

NEWRY, (893), l. in 1881; br. J. E. Farmer, s Goliah 2nd (613), d by Carouser (341), g d by Union Jack (1252).

NIGHTINGALE'S No. 1, 1873 (894), l. in 1872; br. V. E. Nightingale, s Son of the Marquis, d by Young Worcester.

NIGHTINGALE'S No. 1, 1880 (895), l. in 1879 ; br. V. E. Nightingale, s Carouser (341), d by a Byrd Ram.

NIGHTINGALE'S No. 2, 1880 (896), l. in 1879 ; br. V. E. Nightingale, s Carouser (341), d by Mansell's No. 2 (808).

NOBLE, (897), l. in 1876; br. J. W. Minton, s Bedford Hero (111), d a Minton Ewe.

NOBLE, (898), l. in 1879 ; br. R. J. Nash, s Nobleman 901), d by Royal Blood (1020).

NOBLE, (899), l. in 1882 ; br. J. Hall, s Richmond (Earl of Zetland), d by Cynic.

NOBLEMAN, (900), l. in 1873 ; br. R. H. Masfen, s The Colonel (Masfen), d by Brother to Standard Bearer (Masfen), g d by Corsair (57), g g d by Corinthian (Coxon).

NOBLEMAN, (901), l. in 1876 ; br. R J. Nash, s Pulley's No. 9, 1874 (978), d by Sheldon's No. 2, 1872, g d a Coxon Ewe.

NOBLEMAN 3rd, (902) ; br. Lord Willoughby de Broke, s Nobleman (37), d a Willoughby de Broke Ewe.

NOCK'S No. 3, 1870 (903), l. in 1869 ; br. T. Nock, s Wellington (P. W. Bowen), d by a Crane Ram.

NOCK'S No. 4, 1870 (904), l. in 1869 ; br. T. Nock, s Shamrock (J. Evans), d by an Evans Ram.

T. NOCK'S No. 5, (905) ; br. T. Nock, s Mansell's 26, 1878.

NOCK'S No. 7, 1881 (906), l. in 1878 ; br. Lord Chesham, s Primate (Evans), d by Lord Vincent (Chesham).

NOCK'S No. 10, (907) ; br. T. Nock, s Bristol Reserve (141).

NONPAREIL, (908), l. in 1864 ; br. J. Evans, s Pride of Pitchford (961). d by Emperor (525), g d by Humphrey Davy (676), g g d by Own Brother to Bossy (925).

NONPAREIL 3rd, (909) ; br. Lord Willoughby de Broke.

NONPAREIL 3rd, (910), l. in 1873 ; br. H. Griffin, s Nonpareil 2nd (Evans), d a Pell Wall Ewe.

NONSUCH, (911), l. in 1879 ; br. T. Nock, s Young Aston (Nock).

NORMANTON, (912), l. in 1877 ; br. S. C Pilgrim, s Lord Radford (773), d by Wanderer (Beach).

NORTH STAR, (913), l. in 1877 ; br. J. Evans, s True Light (1242), d by Grand Duke (620), g d by Royalty (1029).

NOTABLE, (914), l. in 1881 ; br. T. Nock, s Jumbo, d by Sandboy (1044).

NUGGET, (916), l. in 1878 ; br. T. Nock, s Lord Aston (H. J. Sheldon).

NUMBER SIX, (917), l. in 1881 ; br F. Gibson, s a Sheldon Ram, d by a Masfen Ram.

ODSTONE, (918), l. in 1874 ; br. W. German, s Confidence (101), d by Commander (415).

ODSTONE, (919), l. in 1879 ; br. Mrs. Barrs, s Corporal (443), d by Major (791).

OLD B., (920), l. in 1849; br. G. Adney, s Buckskin (Adney).

OMAR PASHA, (921), l. in 1881; br. J. Darling, s Dudmaston Hero (165), d by Grand Turk (629).

ONEY, (922), l. in 1878; br. Mr. F. Bach, s Tartar, (1173), d by The Warden (1210).

ORANGEMORE, (923), l. in 1876; br. J. Coxon, s Champion (351), d by Captivator (76), g d by Preserver (950).

OUR CHIEF, (924), l. in 1871; br. C. R. Keeling, s Commander-in-Chief (416), d by Pride of Pendeford (Masfen).

OWN BROTHER TO BOSSY, (925), l. in 1850; br. S. Meire, s Perfection (Meire), d by Old Profit (Meire).

OXFORD HERO, (926), l. in 1880; br. J. W. Minton, s Son of Coxcomb (Crane), d by Bedford Hero (111).

OXON CALIGULA, (927), l. in 1878; br. R. Edwards, s Caligula 2nd (323), d by Young Lord Warden (Edwards).

PACKINGTON DUKE, (928), l. in 1862; br. Earl of Aylesford, s an Adney Ram.

PADDY, (929), l. in 1868; br. J. & E. Crane, s Shamrock (1054), d by Chieftain (384).

PADDY GREEN, (930), l. in 1872; br. J. Evans, s Broadgauge (89), d by Pride of Pitchford (961), g d by Humphrey Davy (676), g g d by Emperor (525).

PATENT, (931), l. in 1858; br. S. Byrd, s Patentee (4), d a Farmer (Shipley) Ewe.

PATENT, (932), l. in 1870; br. T. Mansell, s Pattern (933), d by Earl of Plymouth (48).

PATTERN (933), l. in 1868; br. T. Mansell, s Conservative (435), d by Big Neck (T. Mansell).

PENSIONER, (935), l. in 1868; br. J. Coxon, s Commander (415), d by Veteran (1255).

PHILISTINE, (936), l. in 1875; br. H. J. Sheldon, s Goliah (611), d a Sheldon Ewe.

PIONEER, (937), l. in 1881; br. Mrs. Barrs, s Lord Oxon (167), d by Lord Beaconsfield (736).

PIRATE, (938), l. in 1867; br. R. H. Masfen, s Corsair (57).

POCKET HERCULES. (939), l. in 1881; br. Viscount Falmouth s Goliath (612), d by Harry (656).

POLICE CONSTABLE, (940), l. in 1881; br. J. L. Naper, s Protector (972), d by Cossack (445).

POPULARITY, (941), l. in 1878; br. J. Coxon, s First Fruits (564), d by Consolation (Coxon), g d by Sheet Anchor (1060).

POSENHALL, (942), l. in 1872; br. Pitt, s Callaughton (Instone), d by Prince Imperial.

POSENHALL, LORDSHIP, (943), l. in 1880; br. G. W. Wheeler, s His Lordship (669), d by Brutus (306), g d by Lord Chesham's No. 19 (364).

POST CAPTAIN, (944), l. in 1877; br. J. Evans, s British Tar (292), d by Union Jack (1252), g d by Nonpareil (908).

PRACTICAL, (945), l. in 1874; br. J. Pulley, s Dorchester Hero (105), d by Proud Salopian (87).

PRECOCITY, (946), l. in 1882; br. T. Nock, s Lord Clive (742), d by Cynic 2nd, (T. Nock)

PREMIER, (947), l. in 1867; br. R. H. Masfen, s Crane's No. 13, 1866, d by Mainstay (Masfen).

PREMIER, (948); br. W. Baker, s Lord Hereford (Pulley).

PREMIER, (949)), l. in 1872; br. J. Coxon, s Captivator (76), d by Sheet Anchor (1060).

PRESERVER, (950); br. Mrs. Wadlow, s P. W. Bowen's No. 11, 1862, d a Wadlow Ewe.

PRETENDER, (951), l. in 1867; br. C. R. Keeling, s Cardinal (53), d by Nuggett, (Keeling).

PRIAM, (952), l. in 1876; br. J. E. Farmer, s Beach's No. 3, 1873 (221), d by Young Heenan.

PRIDE OF BASCHURCH, (953), l. in 1880; br. R. Thomas, s A1 (176), d by Prince (963), g d by Calcot Chieftain (319).

PRIDE OF BISHTON, (954), l. in 1877; [br. M. Williams, s Bristol Reserve (144), d by Co-Monument (422), g d by Exile (540).

PRIDE OF BREWOOD, (955), l. in 1882; br. J. Beach, s Minton's Pride (858), d by Snowflake (1120).

PRIDE OF DUDMASTON, (956), l. in 1881; br. T. J. Mansell, s Pride of Montford (959), d by Double B. (500).

PRIDE OF FREEFORD, (957), l. in 1880; br. J. Coxon, s City Member (391), d by Marquis of Bath (822), g d by Preserver (950).

PRIDE OF HATTON'S, (958), l. in 1882; br. J. Beach, s Minton's Pride (858), d by Challenge (347).

PRIDE OF MONTFORD, (959), l in 1877; br. T. S. Minton, s Son of Calcot (1125), d a Minton Ewe.

PRIDE OF OXON, (960), l. in 1879; br. R. Edwards, s Caligula 2nd (322), d by Comrade.

PRIDE OF PITCHFORD, (961), l. in 1860; br. W. P. Claridge, s Young Crane (Claridge), d by Grey Friar (Adney), g d by Brother to Duke of Gloster (Horton), g g d by Old Rowton (Lee).

PRIMUS, (962), l. in 1880; br. J. Harding, s Caligula 2nd (322), d by Sheldon's No. 3, 1876 (1063), g d by Marquis (820).

PRINCE, (963), l. in 1874; br. J. & E. Crane, s Claudius (103), d by Celebrity (6).

PRINCE, (964), l. in 1881; br. J. Harding, s. Caligula 2nd (322), d by His Lordship (669).

PRINCE CHARLIE, (965), l. in 1874; br. J. Evans, s Grand Duke (620), d by Hardlines (647), g d by Pride of Pitchford (961).

PRINCE IMPERIAL, (966), l. in 1875; br. Mrs. S. Beach, s British Oak (290), d by Latimer (H. Smith).

PRINCE KNIGHTLEY, (967), l. in 1878; br. G. Allen, s Lord Knightley (T. Mansell), d by Black Prince.

PRINCE LEWELLIN, (968), l. in 1881; br. T. Fenn, s Caligula 2nd (322), d by Ensdon Hero (104), g d by Midlothian (842), g g d by Rob Roy (1018).

PRINCE OF WALES, (969), l. in 1880; br. R. Thomas, s Grand Chief (618), d by Prince (963), g d by Benthall Chieftain (246).

PRINCE REGENT, (970), l. in 1881; br. J. L. Naper, s Protector (972), d by Longbow (728).

PRINCE ROYAL, (971), l. in 1880; br. Mr. T. S. Minton, s Royal Reserve (159), d a Matthews Ewe.

PROTECTOR, (972), l. in 1879; br. R. H. Masfen, s Philistine (R. H. Masfen), d by Columbus (128), g d by Coxswain (R. H. Masfen), g g d by True Type (85).

PROTECTOR, (973), l. in 1881; br. R. Thomas, s Prince Victor (158), d by Grandeur (133), g d by Benthall Chieftain (246).

PROTOTYPE, (974), l. in 1881; br. J. L. Naper, s Protector (972), d by Longbow (728).

PULLEY, (975), l. in 1880; br. Capt. J. B. Haydock, s Wootton (1279), d by Blood Royal.

PULLEY RAM, (976), l. in 1871; br. J. Pulley, s Dorchester Hero (105).

PULLEY RAM, (977), l. in 1872; br. J. Pulley, s Proud Salopian (87).

PULLEY'S No. 9, 1874 (978), l. in 1873; br. J. Pulley, s Fat Back Patentee (Pulley), d by Radford (989).

PULLEY'S No. 1, 1876 (979), l. in 1875 ; br. J. Pulley, s Sultan (1167), d by Buckskin (307).

PULLEY'S No. 1, 1882 (980), l. in 1879 ; br. J. Pulley, s Young Colossus (1301), d by Young Buckskin.

PUNCH, (981), l. in 1880 ; br. J. Beach, s Seaman (1047), d by Lord Vincent (Chesham).

QUALITY, (982), l. in 1877 ; br. J. Coxon, s Courtier (454), d by Stamina (1157), g d by Preserver (950).

QUALITY, (983), l. in 1878 ; br. G. German, s Sir Robert (1108), d by Confidence (101).

QUALITY, (984), l. in 1880 ; br. T. S. Minton, s Marquis of Bath (822), d by Dudmaston (504).

QUALITY 3rd, (985), l. in 1864 ; br. S. Byrd, s Quality, d by Patentee (4).

QUARRYMAN, (986), l. in 1878 ; br. M. Williams, s Bristol Reserve (!44), d by Co Monument (422), g d by Exile (540).

QUICKSTEP, (987), l. in 1868 ; br. The late J. Beach, s Cardinal (53), d a Beach Ewe.

RANDELL'S No. 1, 1876 (988), l. in 1875 ; br. C. Randell, s Our Chief (924), d by Pride of Pendeford (Masfen).

RADFORD, (989), l. in 1867 ; br. H. Smith, s Rejected (H. Smith, d by Perfection (Foster).

RADICAL, (990), l. in 1871 ; br. W. German, s Commodore (418), d by Novelty (41).

RAM LAMB, (991), in 1882 ; br. W. German, s Robin Rough (1016), d by Lord of the Manor (771).

RANGER, (992), l. in 1872 ; br. J. Coxon, s Stamina (757), d by Commander (415).

READY MONEY, (993), l. in 1880 ; br. T. S. Minton, s Son of Coxcomb (Crane), d by Foreman (568).

REALITY, (994), l. in 1880 ; br. J. Evans, s Royal Taunton (115), d by Broadgauge (89).

RECTOR, (995), l. in 1874 ; br. W. German, s Crane's No. 11, 1873 (468), d by Conductor (90).

REDDITCH WONDER, (996), l. in 1882 ; br. E. M. Coleman, s Mansell's No. 12, 1881 (811), d a Minton Ewe.

REDNAL CHIEF, (997), l. in 1879 ; br. R. Meredith, s Chief (383), d by The Tory (1208).

REFLECTION, (998), l. in 1872 ; br. Lord Chesham, s Oxford Hero (77), d a Chesham Ewe.

REFORMER, (999), l. in 1873 ; br. Lord Chesham, s Son of Oxford Hero (1142), d by Latimer (H. Smith).

REGULATOR, (1000), l. in 1878; br. J. Coxon, s Marquis of Bath (822), d by Confidence (101), g d by Apollo (190).

REINDEER, (1001), l. in 1869 ; br. R. H. Masfen, s Compton Buck (A. H. Minor), d by Marrowfat (Masfen).

RELIANCE, (1002), l. in 1881 ; br. T. S. Minton, s Ercall P. (528), d by Son of Bedford Hero (1123).

RESTORER, (1003), l. in 1879 ; br. J. Evans, s Royal Taunton (115), d by Grand Duke (620), g d by Hardlines (647).

RICHMOND, (1004), l. in 1878 ; br. J. Beach, s Sir Garnet (1091), d by Lord Stafford (775).

RICHMOND, (1005). l. in 1880) ; br. J. Evans, s Royal Taunton (115).

RIDLEY, (1006), l. in 1877 ; br. G. German, s Courtier (454), d by Crown Prince (483).

RIFLEMAN, (1007), l. in 1868; br. E. Thornton, s Volunteer (51), d by Canterbury Patentee (15).

RIFLEMAN, (1008), l. in 1869 ; br. W. German, s Young Volunteer (1330), d by Young Viscount.

RIFLEMAN, (1009), l. in 1881 ; br. R. V. C. Groves, s Mark Anthony (819) ; d by Volunteer 2nd (1261).

RIGHTON, (1010), l. in 1874 ; br. W. German, s Young Claudius (1297), d by Chancellor (353).

R.N., (1011), l. in 1877 ; br. Mrs. S. Beach, s British Oak (290), d by Cardigan (338), g d by Young Gayton (1307).

ROBERTSON, (1012), l. in 1876 ; br. Viscount Falmouth, s Harry (656), d by Jemmy (686), g d by Yates No. 6 (1285).

ROBIN, (1013), l. in 1875 ; br. W. German, s Confidence (101), d by Young Volunteer (1330).

ROBIN HOOD, (1014), l. in 1880 ; br. C. Randell, s British Chieftain (288), d by Colossal (410).

ROBIN HOOD 2nd, (1015), l. in 1880 ; br. Lord Chesham, s Lord Chesham's No. 3.

ROBIN ROUGH, (1016), l. in 1880 ; br. T. J. Mansell, s Pride of Montford (959), d by Double B (500).

ROBIN ROUGH NOB, (1017), l. in 1874 or 5 ; br. G. Allen, s Commander-in-Chief 2nd, d by Fat Back (551), g d by Lord Clifden, g g d by Canterbury Patentee (15), g g g d by Patentee (4).

ROB ROY, (1018), l. in 1867; br. S. Byrd, s Breeders Friend (S. Byrd), d by Lord Clifden (44), g d by Criterion.

ROGUE, (1019), l. in 1876; br. G. German, s Conserver (437), d by Consolation (438).

ROYAL BLOOD, (1020), l. in 1871; br. T. Nock, s Blood Royal (83).

ROYAL CARDINAL, (1021), l. in 1881; br. J. Evans, s Royal Taunton (115), d by Royalty (1029), g d by Abbot of Bury (54).

ROYAL CHIEF, (1022), l. in 1879; br. J. Evans, s Royal Taunton (115), d by British Oak (290), g d by Proud Salopian (87).

ROYAL ENSIGN, (1023), l. in 1881; br. Lord Chesham, s Dudmaston (506), d by Royal Aston (124).

ROYAL GEM, (1024), l. in 1880; br. J. Evans, s Royal Taunton (115), d by British Oak (290), g d by Lord Uffington (56), g g d by Competition (425).

ROYAL GRAND DUKE, (1025), l. in 1880; br. J. Evans, s Royal Taunton (115), d by Grand Duke (620), g d by Hardlines (647).

ROYAL LEGATEE, (1026), l. in 1872; br. C. Byrd, s Legatee (95), d by Blood Royal (83).

ROYAL RESERVE, (1027), l. in 1875; br. R. J. Nash, s Sheldon's No. 2, 1872 (1062), d a Fota Ewe.

ROYAL STANDARD, (1028), l. in 1881; br. J. Evans, s Royal Taunton (115), d by British Oak (290), g d by Union Jack (1252).

ROYALTY, (1029), l. in 1669; br. C. R. Keeling, s Cardinal (53), d by Competition (425).

ROYAL VICTOR, (1030), l. in 1879; br. J. Evans, s Royal Taunton (115), d by British Oak (290), g d by Union Jack (1252).

ROWLAND, (1031), l. in 1874: br. W. German, s Neptune (890), d by Rifleman (1008).

RUFUS, (1032), l. in 1873; br. W. German, s Crown Prince (483), d by Captivator (76).

RUGBY, (1033), l. in 1880; br. Capt. J. B. Haydock, s Wootton 1279, d by Blood Royal.

RULER, (1034), l. in 1879; br. G. German, s Farmer's Friend (548), d by Champion (351).

SAFEGUARD, (1035), l. in (1869); br. W. German, s Preserver (950), d by Commander (415).

SAFEGUARD, (1036), l. in 1882; br. R. Thomas, s The Patriot (1198), d by Prince (963), g d by Co-Monument (423).

SAFETY VALVE, (1037), l. in 1881; br. T. S. Minton, s Montford Hero (160), d by Son of Conservative (1134).

SALISBURY, (1038), l. in 1877; br. W. H. Clare, s Dudmaston (504)), d by Diamond.

SALOPIAN C, (1039), l. in 1880; br. Crane & Tanner, s Dudmaston (504), d by Salopian.

SAM, (1040), l. in 1871; br. Viscount Falmouth, s Yates, No. 6, 1870 (1285).

SAM, (1041), l. in 1879; br. The Earl of Shrewsbury, s. Minton's No. 4, 1877, d by Bedford Hero (111).

SAMPSON, (1042), l. in 1865; br. S. Byrd, s Prince (Byrd), d by Constitution (C. Byrd).

SAMSON, (1043); br. Viscount Falmouth, s Yates No. 6, 1870 (1285), d by Nock's No. 4, 1870 (904).

SANDBOY, (1044), l. in 1874; br. R. H. Masfen, s His Majesty (T. Mansell), d by Commander-in-Chief (416), g d by Corsair (57), g g d by Corinthian (Masfen).

SCOTTISH CHIEF, (1045), l. in 1871; br. R. H. Masfen, s Commander-in-Chief (416), d by Grindle (Keeling), g d by Havelock (27), g g d by Brother to Gratitude (R. H. Masfen).

SCOTTISH HERO, (1046), l. in 1868; br. John Evans, s Nonpareil (908), d by Competition (425).

SEAMAN, (1047), l. in 1878; br. Crane & Tanner, s Dudmaston (504), d by Chivalry (378).

SECOND TO NONE, (1048), l. in 1873; br. Lord Chesham, s Son of Oxford Hero, (1142), d by Latimer (H. Smith).

SECUNDUS, (1049), l. in 1873; br. R. H. Masfen, s Rob Roy (1018), d by Grindle (Keeling), g d by Marrowfat (Masfen), g g d by Lady Jane Grey's Son (Masfen).

SEVERN, (1050), l. in 1870 or 1871; br. H. Matthews, s The Rejected (T. Mansell), d a Matthews Ewe.

SEVERUS 2nd, (1051), l. in 1874; br. W. F. Firmstone, s Severus (T. J. Mansell), d by Crosswood Hero 2nd.

SEVERUS 2nd, (1052), l. in 1877; br. J. W. Minton, s Severus (T. J. Mansell), d a Minton Ewe.

SEXTUS, (1053), l. in 1881; br. Lord Chesham, s Dudmaston (506), d by Mrs. Beach's No. 1, 1877.

SHAMROCK, (1054), l. in 1865; br. C. W. Hamilton, s Emperor d by Garry Owen.

SHAMROCK, (1055), l. in 1876 ; br. T. Mansell s Raby Duke (108), d by Landseer (700).

SHELDON'S A, (1056), l. in 1881 ; br. H. J. Sheldon, s Lord Blaydon (738), d a Sheldon Ewe.

SHELDON'S B, (1057), l. in 1881 ; br. H. J. Sheldon, s Beach's No. 3, 1880, d a Sheldon Ewe.

SHELDON'S C, (1058), l. in 1881 ; br. H. J. Sheldon, s Beach's No. 3, 1880, d a Sheldon Ewe.

SHELDON'S D, (1059), l. in 1881 ; br. H. J. Sheldon, s Graham's No. 6, 1880 (614), d a Sheldon Ewe.

SHEET ANCHOR, (1060), l. in 1865 ; br. J. Coxon, s Duke of Newcastle (39) ; d by Nobleman (37,) g d by Veteran (1255).

SHELDON'S KILBURN RAM, (1061), l. in 1878, br. H. J. Sheldon, s Philistine (936), d a Sheldon Ewe.

SHELDON'S No. 2, 1872 (1062), l. in 1871 ; br. H. J. Sheldon, s Horley's No. 29, 1870 (673).

SHELDON'S No. 3, 1876 (1063), l. in 1874, br. H. J. Sheldon, s Goliah (611), d a Sheldon Ewe.

SHELDON'S No. 4, 1876 (1064), l. in 1876 ; br. H. J. Sheldon, s Taunton Reserve (118), d a Sheldon Ewe.

SHELDON'S No. 4, 1879 (1065), l. in 1878 ; br. H. J. Sheldon, s Sheldon's No 1.

SHELDON'S No. 2, 1881 (1066), l in 1880 ; br. H. J. Sheldon, s Model (862).

SHELDON'S RAM, (1067) l. in 1861 ; br. H. J. Sheldon.

SHELDON'S RAM, (1068), l. in 1873 ; br. H. J. Sheldon.

SHELDON'S RAM, (1069), l. in 1875 ; br. H. J. Sheldon.

SHELDON'S RAM, 1879 (1070), l. in 1878 ; br. H. J. Sheldon.

SHELDON'S RAM, (1071) ; br. H. J. Sheldon, s Goliah (611).

SHELFORD, (1072), l. in 1878 ; br. S. C. Pilgrim, s Lord Melton (Pilgrim), d by Commandant (Keeling).

SHERIFF, (1073), l. in 1872 ; br. Viscount Falmouth, s Jemmy (686), d a Falmouth Ewe.

SHIPSTON, (1074), l. in 1875 ; br. H. J. Sheldon, s Goliah (611), d a Sheldon Ewe.

SHIPSTON, (1075), l. in 1877 ; br. H. J. Sheldon, s Taunton Reserve (118).

SHORT-LEGGED PATENTEE, (1076) ; br. Earl of Dartmouth, s Patentee (4), d a Dartmouth Ewe.

SHRAWARDINE, (1077), l. in 1867 ; br. P. W. Bowen.

SHRAWARDINE, (1078), l. in 1878; br. Crane & Tanner, s Dudmaston (504), d by Caligula (112).

SHRAWARDINE, (1079), l. in 1879; br. Crane & Tanner, s Dudmaston (504), d by Claudius (103).

SHRAWARDINE, (1080), l. in 1879; br. Crane & Tanner, s Bristol Reserve (144), d by Claudius (1C3).

SHREWSBURY, (1081), l. in 1873; br. Lord Chesham, s Duke of Manchester (70), d a Crane Ewe.

SHREWSBURY HERO, (1082), l. in 1879; br. J. W. Minton, s Marquis of Bath (822), d by Son of Calcot (1125).

SHROPSHIRE RAM (1083), l. in 1836; br. T. L. Meire, s a Meire Ram, d a Meire Ewe.

SHROPSHIRE RAM, (1084), l. in 1850; br. S. Meire, s a Meire Ram.

SHROPSHIRE RAM, (1085), l. in 1874; br. Mr. Bostock.

SHROPSHIRE RAM, (1086), l. in 1878; br. T. L. Meire, s a Meire Ram, d a Meire Ewe.

SHROPSHIRE RAM, (1087), l. in 1880; br. Sir W. W. Wynn, Bart.; s Lord Ordsley (Mrs. Smith), d an Evans Ewe.

SHROPSHIRE RAM, (1088), l. in 1881; br. T. Mansell, s Lord Clive (742), d by Milton (843).

SIMON DE MONTFORD, (1089), l. in 1880; br. J. Harding, s Pride of Montford (959), d by Mansell's No. 16, 1876 (802).

SIR EVELYN, (1090), l. in 1880; br. J. Harding, s Pride of Montford (959), d by Mansell's No 16, 1876 (802), g d by Marquis (820).

SIR GARNET, 1091, l. in 1877; br. Mrs. S. Beach, s Masterman (828), d by Duke of Manchester (70).

SIR GARNET, (1092), l. in 1879; br. T. J. Mansell, s Pride of Montford (959), d a Wadlow Ewe.

SIR GEORGE, (1093), l. in 1878; br. Capt. H. Townsend, s Freeman (585), d by Baron (210).

SIR GEORGE, (1094), l. in 1880; br. J. L. Naper, s Sir Guy (147), d by Conway (440).

SIR GEORGE, (1095), l. in 1881; br. G. German, s Lord Oxon (167), d a German Ewe.

SIR GRAY, (1096), l. in 1873; br. J. L. Naper, s Viceroy (1256), d by Speculum (1154).

SIR GREGORY, (1097), l. in 1881; br. J. L. Naper, s Sir Guy (147), d by Quality (982).

SIR HARRY, (1098), l. in 1861 ; br. H. Smith, s Viscount, d by Perfection.

SIR HARRY, (1099), l. in 1879 ; br. Mrs. Smith, s Multum-in-Parvo (885), d by Ensdon Hero (104).

SIR HENRY, (1100), l. in 1879 ; br. H. Griffin, s Lord Chesham's No. 13, 1877, d by Mansell's No. 8.

SIR JOSEPH, (1101), l. in 1876 ; br. Mrs. S. Beach, s Masterman (828), d by Duke of Manchester (70), g d by Cardinal (53).

SIR JOSEPH, (1102), l. in 1880 ; br. Lord Chesham, s Young Colossus, d by Royal Taunton (115).

SIR JOSEPH, (1103), l. in 1881 ; br. J. Pulley, s Young Colossus (1301), d by Worcester.

SIR MATTHEW, (1104), l. in 1880 ; br. M. Williams, s Hercules (663), d by Co-Monument (423).

SIR PROBERT, (1105), l. in 1881 ; br. H. J. Sheldon, s G. Graham's No. 6, 1880 (614), d by Prince Imperial (966).

SIR RICHARD, (1106), l. in 1879 ; br. T. Mansell, s Raby Duke (108), d by Landseer (700).

SIR ROBERT, (1107), l. in 1873 ; br. Viscount Falmouth, s Harry (656), d by Jemmy (686).

SIR ROBERT, (1108), l. in 1877 ; br. R. H. Masfen, s Hero, d by Commander in-Chief (416).

SIR ROGER, (1109), l. in 1871 ; br. Mrs. S. Beach, s Duke of Manchester (70), d a Beach Ewe.

SIR ROGER, (1110), l. in 1873 ; br. Viscount Falmouth, s Yates No. 6, 1870 (1285), d a Bradburne Ewe.

SIR ROGER, (1111), l. in 1878 ; br. B. Rogers, s a Matthews Ram, d by a Crane and Tanner Ram, g d by a Masfen Ram.

SIR ROGER, (1112), l. in 1878 ; br. Lord Chesham, s Mansell's No. 6, 1875, d by Lord Kingston (99), g d by Milton.

SIR SAMUEL, (1113), l. in 1860 ; br. S. Meire, s Magnet (Meire), d by Profit 2nd (Meire), g d by Magnum Bonum (3).

SIR WILLIAM, (1114), l. in 1874 ; br. W. Baker, s Tarquin (R. H. Masfen), d by Superior.

SIR WILLIAM, (1115), l. in 1877 ; br. J. Coxon, s Courtier (455), d by Stamina (1157).

SIR WILLIAM, (1116), l. in 1879 ; br. T. Mansell, s County Member (452), d by Rifleman (1007).

H. SMITH'S No. 7, (1117), br. Henry Smith.

MRS. SMITH'S No. 12, 1875 (1118), l. in 1874 ; br. Mrs. Smith, s General Benbow (Mrs. Smith).

SMITH RAM, (1119), l. in 1876 ; br. Mrs. Smith.

SNOWFLAKE, (1120), l. in 1877 ; br. Mrs. S. Beach, s Manager (795), d by Latimer (H. Smith), g d by Gayton Prince (592).

SNOWFLIGHT, (1121), l. in 1878 ; br. J. Beach, s Snowflake (1120), d by Masterman (828).

SON OF BACH'S RAM, (1122), l. in 1859 ; br. B. Vaughan, s Bach's Ram (203), d a Burway Ewe.

SON OF BEDFORD HERO, (1123), l. in 1875 ; br. J. W. Minton s Bedford Hero (111), d a Minton Ewe.

SON OF BRISTOL RESERVE, (1124), l. in 1878 ; br. M. Williams, s Bristol Reserve (144), d by Lord Clifden (44).

SON OF CALCOT, (1125), l. in 1874 ; br. T. Mansell, s Calcot (317), d a Mansell Ewe.

SON OF CARDINAL, (1126), l. in 1873 ; br. J. Evans, s Cardinal (53).

SON OF CHAMPION (1127), l. in 1876 ; br. G. German, s Champion (351), d by Confidence (101).

SON OF CHIVALRY, (1128), l. in 1875 ; br. E. Crane, s Chivalry (387), d by Claudius (103).

SON OF CLAUDIUS, (1129) ; br. B. Walker, s Claudius (103), d by Blood Royal

SON OF CLAUDIUS, (1130), l. in 1873 ; br. T. Fenn, s Claudius (103), d by Shamrock (1054).

SON OF COLUMBUS, (1131), l. in 1878 ; br. Crane & Tanner, s Columbus (128), d by Chivalry (387)

SON OF COMMANDER-IN-CHIEF, (1132), l. in 1873 ; br. T. Horley, s Commander-in-Chief (416), d by Carmelite.

SON OF CONSERVATIVE, (1133), l. in 1869 ; br. T. Mansell, s Conservative (435), d by Marquis (820).

SON OF CONSERVATIVE, (1134), l. in 1872 ; br. T. Mansell, s Conservative (435), d by P. W. Bowen's, No. 11, 1862.

SON OF HENRY SMITH, (1135), l. in 1861 ; br. V. E. Nightingale, s Henry Smith (661), d a Nightingale Ewe.

SON OF HERO, (1136), l. in 1877 ; br. H. Lowe, s Hero (668), d by Utility (1253). .

SON OF ISLANDER, (1137), l. in 1872 ; br. D. R. Davies, s Islander.

SON OF LEVIATHAN, (1138) ; br. Lord Willoughby de Broke, s Leviathan (72).

SON OF LITTLE LORD, (1139), l. in 1872; br. T. Mansell, s Little Lord (718), d by Conservative (435), g d by Earl Plymouth (48).

SON OF LORD CARLISLE, (1140), l. in 1881; br. Peter Everall, s Lord Carlisle (163), d a Matthews Ewe.

SON OF LORD CLIFDEN, (1141, l. in 1865; br. T. Mansell, s Lord Clifden (44), d by Mansell's No. 6, 1859 (797).

SON OF OXFORD HERO, (1142), l. in 1874; br. Lord Chesham, s Oxford Hero (77), d by Matthew (79).

SON OF PATTERN, (1143), l. in 1874; br. Lord Chesham, s Pattern (933), d by Masterman (828).

SON OF PRINCE, (1145), l. in 1878; br. R. Thomas, s Prince (963), d by Co-Monument (423), g d by The Ruler (1204).

SON OF QUALITY, (1146); br. Lord Willoughby de Broke, s Quality 3rd (985).

SON OF STANDARD BEARER, (1147), l. in 1881; br. The Earl of Strathmore, s Standard Bearer (80), d by Crane's No. 2.

SON OF TRUE TYPE, (1148), l. in 1873; br. C. R. Keeling, s True Type (85), d a Keeling Ewe.

SON OF WORCESTER, (1149). l. in 1866; br. J. Hand, s Worcester (1280), d a Wigley Ewe.

SOUTHPORT, (1150), l. in 1866; br. H. Smith, s Rejected (H. Smith).

SOUTHPORT PRIZE, (1151), l. in 1880; br. J. Beach, s. First Choice (561), d by Cardigan (338).

SPECULATION (1152), l. in 1878; br. T. J. Mansell, s Shamrock, (1055), d by True Stock (1243).

SPECULATOR (1153), l. in 1876; br. R. H. Masfen, s Capitalist (331), d by The Earl (Dartmouth).

SPECULUM (1154), l. in 1867; br. T. Horley, s Duke of Newcastle (39), d by Hagley.

STAFFORD (1155), l. in 1877; br. Lord Chesham, s Royal Taunton (115), d by Sir Charles (C. Byrd), g d by Latimer (H. Smith).

STAFFORD HERO (1156), l. in 1880; br. Lord Chesham, s Mrs. Beach's No. 1, 1877, d by British Oak (290), g d by Duke of Manchester (70).

STAMINA (1157), l. in 1870; br. T. Mansell, s Rifleman (1007), d by Conservative (435).

STANDARD (1158), l. in 1870; br. J. Evans, s Standard Bearer (80), d by Nonpareil (908).

STANDARD (1159), l. in 1879 ; br. R. H. Masfen, s Clinker (397) d by His Majesty (T. Mansell).

STAR, (1160), l. in 1880 ; br. T. Mansell, s North Star (913), d by Raby Duke (108).

STATESMAN (1161), l. in 1877 ; br. H. J. Sheldon, s Beach's No. 9, 1876; (224), d a Sheldon Ewe.

STERLING, (1162), l. in 1878 ; br J. Coxon, s Sir Robert (1108), d by Colossus (411).

STORMER, (1163), l. in 1878 ; br. T. Fenn, s Ensdon Hero (104), d by Son of My Lord, g d by Kingscraft (698).

STRAIGHT BACK, (1164), l. in 1870 ; br. J. Evans.

STUDLEY HERO, (1165), l. in 1882 ; br. E. M. Coleman, s Mansell's No. 12, 1881 (811), d a Minton Ewe.

SULPHUR, (1166), l. in 1881 ; br. T. Mansell, s 3rd Marquis of Bute (132), d by True Stock (1243).

SULTAN, (1167), l. in 1872 ; br. Lord Chesham, s Harrogate (Lord Chesham).

SULTAN 3rd, (1168), l. in 1881 ; br. J. Pulley, s Young Sultan (117), d by Colossus (411).

SUTTON MADDOCK, (1169), l. in 1878 ; br. Mrs. H. Smith, s Belligerent (T. Nock), d by Son of Cardinal.

SWEET WILLIAM, (1170) ; br. G. A. May, s Picture, d by Patent (931).

SYMMETRY, (1171), l. in 1881 ; br. T. Mansell, s North Star (913), d by Raby Duke (108).

TAMWORTH, (1172), l. in 1868 ; br. T. Nock, s Chieftain (Nock).

TARTAR, (1173), l. in 1875 ; br. R. H. Masfen, s The Marshall (Masfen), d by True Type (85), g d by Commander-in-Chief (416).

TARTAR, (1174), l. in 1878 ; br. Hy. Townshend, s Freeman (585), d a Beach Ewe.

TAUNTON DUKE, (1175), l. in 1880 ; br. J. Evans, s Royal Taunton (115), d by Grand Duke (620).

TERN, (1176), l. in 1853 ; br. Mr. Juckes, s Son of Farmer, (Adney) d a Juckes Ewe.

THE BARON, (1177), l. in 1875 ; br. Lord Chesham, s Young Harrogate (Chesham), d by Radford (989).

THE BUCK, (1178), l. in 1869 ; br. T. Mansell, s Son of Conservative (1134), d by Tame Deer (T. Mansell).

THE COLONEL, (1179), l. in 1876 ; br. Viscount Falmouth, s Happy-go-Lucky (645), d by Bachelor (199).

THE COLONEL, (1180), l. in 1881 ; br. G. German, s Lord Oxon (167)), d by Courtier (454).

THE COUNCILLOR, (1181), l. in 1881 ; br. J. Coxon, s Beaudesert (239), d by Colossus (411).

THE CZAR, (1182), l. in 1875 ; br. G. German, s Colossus (411), d by Confidence (101).

THE EARL, (1183) ; br. Lord Willoughby de Broke, s Young Patentee (1320).

THE EARL, (1184), l. in 1879 ; br. T. J. Mansell, s May Duke (837), d by Calcot (317), g d by Rifleman (1007).

THE EARL'S No. 14, 1882 (1185), l. in 1881 ; br. The Earl of Shrewsbury, s Grand Seignor (626), d by Scottish Chief (1045).

THE GENERAL, (1186) ; br. Mr. T. Mansell, s Rifleman (1007), d by Conservative (435).

THE GRECIAN, (1187), l. in 1875 ; br. H. Griffin, s Raby Duke (108).

THE HERO, (1188) ; s Captor (334), d by Oxford Hero (77).

THE HERO, (1189), l. in 1882 ; br. T. Nock, s Jumbo (T. Nock), d by T. J. Mansell's No. 35.

THE KNIGHT, (1190), l. in 1870 ; br. T. Mansell, s Conservative (435), d by Mansell, No. 36, 1859 (797).

THE KNIGHT, (1191), l. in 1871 ; br. R. H. Masfen. s Commander (415), d by Black Prince 2nd (26), g d by Mainstay (R. H. Masfen), g g d by Lady Jane Grey's Son (Masfen).

THE LAIRD, (1193), l. in 1875 ; br. T. Fenn, s Midlothian (842), d by Lord Kenyon.

THE MODEL, (1194), l. in 1863 ; br. S. Byrd, s Monarch (866), d by Patentee (4).

THE MOOR, (1195), l. in 1874 ; br. W. Baker, s Tarquin (R. H. Masfen), d by a Byrd Ram.

THE NOBLE, (1196) l. in 1881 ; br. by R. Thomas, s Grand Master (621), d Foreman (568), g d Castle Warden (342).

THE NORMAN, (1197) ; s William the Conqueror.

THE PATRIOT, (1198), l. in 1880 ; br. T. J. Mansell, s Pride of Montford (959), d by May Duke (837).

THE PATRIOT LORD, (1199), l. in 1881 ; br. C. Wadlow, s Bridgnorth (274), d by Warwick (1268).

THE PEER, (1200) ; br. Hon. Noel Hill, s Patron, d by Defiance (Horton).

THE POET, (1201), l. in 1831 ; br. T. S. Minton, s Milton (843), d by Marquis of Bath (822).

THE PRIOR, (1202) ; br. G. German.

THE PROCTOR, (1203), l. in. 1872 ; br. C. Byrd ; s Oxford Hero (77), d by Lord Napier (764).

THE RULER, (1204), l. in 1862 ; br. J. Evans, s Emperor (525), d by Humphrey Davy (676).

THE RULER, (1205), l. in 1878 ; br. B. Walker, s Lord Odstone (766), d by Claudius (103).

THE STAR, (1206), l. in 1877 ; br. R. H. Masfen, s The Gem (Masfen), d by True Type (85), g d by Corsair (57), g g d by Old Crop (Masfen).

THE SULTAN, (1207), l. in 1881 ; br. E. Johnson, s Sultan of Farringdon (Pulley), d g Blood Royal 2nd (Mrs. Smith), g d by The Shah, g g d a Pulley Ewe.

THE TORY, (1208), l. in 1872 ; br. T. Mansell, s Calcot (317), d by Conservative (435).

THE VICTOR, (1209), l. in 1881) ; br. R. Thomas, s Prince Victor (R. Thomas), d by Grandeur (133), g d by Benthall Chieftain (246).

THE WARDEN, (1210), l. in 1869 ; br. Andrews, s Lord Warden (777), d by Lord Clifden (44 '.

THE YOUNG DUKE, (1211) ; br. Thomas Horton, s Duke of Kent (13), d by Old Salop.

THE YOUNG SULTAN, (1212), l. in 1875 ; br. J. Pulley, s Sultan, (1167), d by Proud Salopian (87).

THICKSET, (1213), l. in 1879 ; br. C. Byrd, s. The Hero (1188), d by Touchstone (1230), g d by Little Lord (718).

THICKSET, *alias* THE LAWYER, (1214), l. in 1880, br. J. Harding, s Pride of Montford (959), d by Mansell's No. 16, 1876 (802), g d by Nonpariel 3rd (910).

THOMAS' No. 17, 1876 (1215), l. in 1879 ; br. R. Thomas, s Claudius 3rd (396), d by Prince (963).

THOMAS' No. 3, 1878, (1216), l. in 1877 ; br. R. Thomas, s Prince (963), d by Foreman (568).

THOMAS' No. 1, 1879 (1217), l. in 1878 ; br. R. Thomas, s Prince (963), d by Calcot Chieftain (319).

THOMAS' No. 2, 1879 (1218), l. in 1878 ; br. R. Thomas, s Grandeur (133), d by Foreman (568).

THOMAS' No. 30, 1879 (1219), l. in 1877 : br. R. Thomas, s Prince (963), d by Co-Monument (423).

THOMAS' No. 19, 1880 (1220), l. in 1879 ; br. R. Thomas, s Lord Ensdon, d by Co-Monument (423), g d by De Broke (494).

THOMAS' No. 4, 1881 (1221), l. in 1880 ; br. R. Thomas, s Grand Chief (618), d by Foreman (568).

THOMAS' No. 4, 1881 (1222), l. in 1879 ; br. R. Thomas, s Grand Chief (618), g d by Co-Monument (423).

THOMAS' No. 19, 1881 (1223), l. in 1880 ; br. R. Thomas, s Grand Chief (618), d by Calcot Chieftain (319), g d by The Ruler (1204).

TORY, (1227) ; br. T. Mansell, s Conservative (435), d by Big Neck (T. Mansell).

TORNADO, (1228), l. in 1881 ; br. J. Evans, s Royal Taunton (115), d by British Oak (290), g d by Broadguage (89).

TORY PEER, (1229), l. in 1869 ; br. T. Mansell, s Conservative (435), d by Marquis (820).

TOUCHSTONE, (1230) ; br. R. H. Masfen, s Capitalist (331), d by True Type (85).

TOWN COUNCILLOR, (1231), l. in 1880 ; br. R. M. Knowles, s Lord Mayor (146), d a Clare Ewe.

TOWNSHEND RAM, (1232), l. in 1874 ; br. Capt. H. Townshend.

TRAVELLER, (1233), l. in 1878 ; br. O. R. Keeling, s the Clipper (R. H. Masfen), d a Firmstone Ewe.

TREBLE C, (1234), l. in 1880 ; br. Crane & Tanner, s Dudmaston (504), d by Columbus (128).

TRIBUNE, (1235), l. in 1878 ; br. C. Byrd, s Touchstone (1230), d Hampton Hero (84), g d by Major (88).

TRICKSTER, (1236), l. in 1877 ; br. C. Byrd, s Touchstone (1230), d by Little Lord (718).

TRIUMPH, (1237), l. in 1879 ; br. W. Baker, s Son of Goliah, d by Tarquin (Masfen).

TRIUNE, (1238), l. in 1878 ; br. C. Byrd, s Touchstone (1230), d Little Lord (718), g d by Guardian (642).

TRUE BLUE, (1239) ; br. Bostock, s Touchstone (1230), d by Monument.

TRUE BORN, (1240), l. in 1872 ; br. R. H. Masfen, s True Type (85), d by Monarch (R. H. Masfen), g d by Brother to Gratitude (Masfen).

TRUE FORM, (1241), l. in 1874 ; br. T. J. Mansell, s True Stock (1243), d by Son of Conservative (1134).

TRUE LIGHT, (1242), l. in 1874 ; br. T. Mansell, s True Stock (1243), d by Conservative (435).

F

TRUE STOCK, (1243), l. in 1872 ; br. C. R. Keeling, s True Type (85), d a Keeling Ewe.

TRUISBE, (1244), l. in 1875 ; br. T. J. Mansell, s True Stock (1243), d by Little Lord (718), g d by Earl of Plymouth (48).

TROOPER, (1245), l. in 1875 ; br. Lord Chesham.

TURK, (1246), l. in 1873 ; br. Viscount Falmouth, s Jemmy (686), d a Yates Ewe.

TYNDALE, (1247), l. in 1881 ; br. J. Evans, s Bristol Reserve (144), d by Royal Grand Duke (1025).

TYPICAL, (1248) ; br. C. Byrd, s Touchstone (1230), d by Double B. (500), g d by Oxford Hero (77).

TYPICAL, (1249), l. in 1872 ; br. C. R. Keeling, s True Type (85), d by Cardinal (53), g d by Grindle (Keeling).

UFFINGTON, (1250), l. in 1870 ; br. J. Evans, s Cardinal (53).

ULTIMATUM, (1251), l. in 1881 ; br. R. M. Knowles, s Multum-in-Parvo (885), d a Coxon Ewe.

UNION JACK, (1252), l. in 1870 ; br. J. Evans, s Standard Bearer (80), d by Cardinal (53), g d by Competition (425), g g d by Young Emperor (1305).

UTILITY, (1253), l. in 1871 ; br. R. H. Masfen, s Mainstay (Masfen), d a Masfen Ewe.

VALIANT, (1254), l. in 1858 ; br. J. Coxon, s Veteran (1255), d by Old Packington.

VETERAN, (1255), l. in 1856 ; br. G. Adney, s Buckskin (Adney), d an Adney Ewe.

VICEROY, (1256), l. in 1872 ; br. J. L. Naper, s Lord Haughton (751), d by Negro (888).

VICTOR 2nd, (1257), l. in 1881 ; br. R. Loder, s Royal Victor (1030), d a Chesham Ewe.

VICTOR 3rd, (1258), l. in 1881 ; br. R. Loder, s Royal Victor (1030), d a Chesham Ewe.

VISCOUNT, (1259) ; br. C. Byrd, s Baronet (C. Byrd), d by Lord Knightley (753).

VICTOR, (1260), l. in 1881 ; br. R. Thomas, s Prince Victor (158), d by Grandeur (133), g d by Son of Conservative (1133).

VOLUNTEER 2nd, (1261), l. in 1866 ; br. The Hon. T. Noel Hill, s Volunteer (51), d a Claridge Ewe.

VULCAN, (1262), l. in 1865 ; br. R. H. Masfen.

WARDEN, (1263), l. in 1877 ; br. R. Thomas, s Foreman (568), d by Castle Warden (342).

WARRIOR, (1264), l. in 1869 ; br. H. Smith, s a Crane Ram.

WARRIOR, (1265), l. in 1879 ; br. R. H. Masfen, s Philistine (936), d by Columbus (128).

WARRIOR, (1266), l. in 1880 ; br. R. Thomas, s Cœur-de-Lion (404), d by Grandeur (133), g d by Calcot Chieftain (319).

WARRIOR, (1267), l. in 1881 ; br. J. E. Farmer, s Carlisle (166), d by Double X (501).

WARWICK, (1268), l. in 1877 ; br. H. J. Sheldon, s Taunton Reserve (118), d a Sheldon Ewe.

WATERPROOF, (1269), l. in 1881 ; br. J. Coxon, s Mackintosh (786), d by Marquis of Bath (822).

WELLINGTON HERO, (1270), l. in 1879 ; br. T. Dicken, s T. J. Mansell's No. 10, 1877 (804), d by True Stock (1243).

WELLINGTON 4th, (1271), l. in 1879 ; br. the late C. W. Hamilton, s Young Wellington, d by Pulley, g d by Old Wellington.

WIGLEY DUKE, (1272), l. in 1878 ; br. J. E. Farmer, s Grand Duke (620), d by a Nightingale Ram

WILLIAMS' No. 1, 1879 (1273), l. in 1878 ; br. M. Williams, s Bristol Reserve (144), d by Claude Duval (394), g d by Claudius (103).

WILLIAMS' No. 5, 1882 (1274), l. in 1881 ; br. M. Williams, s Hercules (663), d by Artist (194).

WILLIAMS' RAM, (1275), l. in 1880 ; br. M. Williams, s Hercules (663).

WILLIAMS' No. 4, 1881 (1276), l. in 1880 ; br. M. Williams, s Pride of Bishton (954), d by Artist (194).

WONDER, (1277), l. in 1877 ; br. T. Fenn, s Beach's No. 17, 1876 (228), d by Ensdon Hero (104), g d by Midlothian (842), g g d by Lord Kenyon.

WONDERFUL, (1278), l. in 1879 ; br. R. Bromley, s Wonder (1277), d by Hardlines (647).

WOOTTON, (1279), l. in 1878 . r. W. H. Clare, s Cato (116), d by Conductor (90).

WORCESTER, (1280) ; br. Lord Wenlock.

WORCESTER PATRON 2nd, (1281), l. in 1886 ; br. P. W. Bowen, s Worcester Patron (35), d a Pryce W. Bowen Ewe.

WRINKLE FACE, (1282), l. in 1868 ; br. Mrs. S. Beach, s Cardinal (53), d a Beach Ewe.

YARDLEY, (1283), l. in 1878 ; br. G. Graham, s Gigantic, d by Patrician (Lord Chesham).

YARDLEY No. 10, 1881 (1284), l. in 1880; br. G. Graham, s Regulator (J. Evans), d by Ram bred by Mr. Graham.

YATES No. 6, 1870 (1285), l. in 1869; br. W. Yates, s Grandson of Patentee.

YATES No. 13, 1877 (1286), l. in 1876; br. W. Yates, s Lord Sutton (Nock), d by Sir Roger (1109).

YEOMAN, (1287), l. in 1878; br. J. E. Farmer, s Lord Acton (731), d by Beach's No. 3, 1873 (221).

YOUNG ALDERMAN, (1288), l. in 1882; br. T. Nock, s The Alderman (Coxon), d by Lord Chesham's No. 2, 1879, g d by T. Nock's No. 5, 1879.

YOUNG BACHELOR, (1289), l. in 1877; br. Viscount Falmouth, s Bachelor (199), d a Bradburne Ewe.

YOUNG BEDFORD HERO, (1290), l. in 1872; br. Earl of Strathmore, s Bedford Hero (111), d by Standard Bearer (80).

YOUNG BUCKSKIN, (1291); br. G. Adney, s Buckskin (Adney), d an Adney Ewe.

YOUNG CALCOT, (1292), l. in 1875; br. T. Mansell, s. Calcot (317), d by Conservative (435).

YOUNG CALIGULA, (1293), l. in 1876; br. E. Crane, s Caligula (112), d a Crane Ewe.

YOUNG CAMBRIAN, (1294), l. in 1872; br. J. Evans, s Cambrian (324), d by Lord Uffington (56), g d by Competition (425), g g d by Quality (32).

YOUNG CARACTACUS, (1295), l. in 1869; br. J. & E. Crane, s Caractacus (335), d by Chieftain (384).

YOUNG CHESTER, (1296), l. in 1859; br. Viscount Falmouth, s Chester (381).

YOUNG CLAUDIUS, (1297), l. in 1871; br. B. Walker, s Claudius (103), d by Young Volunteer (1330).

YOUNG CLINKER, (1298), l. in 1878; br. J. H. Bradburne, s Clinker (397), d by Marquis of Bute (91).

YOUNG CLIFDEN, (1299), l. in 1863; br. T. Mansell, s. Lord Clifden (44), d by Short-legged Patentee (1076).

YOUNG COCKADE, (1300), l. in 1881; br. J. Coxon, s Cockade (400), d by Magnet (788).

YOUNG COLOSSUS, (1301), l. in 1877; br. Mrs. S. Beach, s Colossus (411), d by The Knight (1190), g d by Latimer (H. Smith).

YOUNG CONSERVATIVE (1302), l. in 1868; br. J. Coxon, s. Conservative (435), d by Duke of Newcastle (39).

YOUNG DUKE, (1303); br. J. & E Crane, s Duke of Newcastle, d a Crane Ewe.

YOUNG DUKE, (1304); br. Lord Willoughby de Broke, s The Young Duke (1211).

YOUNG EMPEROR, (1305), l. in 1860; br. J. Evans, s Emperor (525), d by Brown Berrington (Meire).

YOUNG ENSDON, (1306), l. in 1873; br. J. Bowen-Jones, s Ensdon Hero (104), d by Lord Warden (777).

YOUNG GAYTON, (1307), l. in 1866; br. The late J. Beach, s Gayton Prince (592), d a Beach Ewe.

YOUNG GENERAL, (1308); br. C. Byrd, s The General (1186), d by Manager (794).

YOUNG GUARDIAN, (1309); br. C. Byrd, s Guardian (642), d by Leese Patentee (S. Byrd).

YOUNG HOPEFUL, (1310), l. in 1879; br. T. Ryland, s Hopeful (671); d a Crane Ewe.

YOUNG KILBURN, (1311), l. in 1879; br. J. Hamar, s Lord Kilburn (149), d by Lord Chesham's No. 4.

YOUNG LORD FALMOUTH, (1312), l. in 1881; br. J. Beach, s Lord Falmouth (746), d an Evans Ewe.

YOUNG MAGNUM BONUM, (1313), l. in 1860; br. S. Meire, s Old Magnum.

YOUNG MANAGER, (1314), l. in 1876; br. J. Beach, s Manager (795), d by Snowflake (1120), g d by Latimer (H. Smith).

YOUNG MARQUIS, (1315), l. in 1867; br. T. Mansell, s Marquis (820), d by Tame Deer (T. Mansell).

YOUNG MASTERMAN, (1316), l. in 1877; br. Mrs. S. Beach, s Masterman, (828), d by Quickstep (987), g d by Latimer (H. Smith).

YOUNG MASTERPIECE, (1317), l. in 1879; br. J. Coxon, s Masterpiece (832), d by Champion (351), g d by Confidence (101).

YOUNG MONTFORD HERO, (1318), l. in 1881; br. T. S. Minton s Montford Hero (160), d by Son of Little Lord (1139).

YOUNG NAPIER, (1319); s Lord Napier (764), d by Criterion (S. Byrd).

YOUNG PATENTEE, (1320), l. in 1863; br. S. Byrd, s Patentee (4), d an S. Byrd Ewe.

YOUNG PERFECTION, (1321), l. in 1865; br. H. Smith, s Lord Harley (Adney), d a Smith Ewe.

YOUNG QUALITY, (1322); br. C. Byrd, s Guardian (642), d by Constitution (S. Byrd).

YOUNG RESERVE, (1323), l. in 1881; br. H. Lee & Son, s Royal Reserve (159), d by Young Caligula (1293).

YOUNG ROYAL, (1324), l. in 1882; br. J. Hamar, s Blood Royal (261), d a Minton Ewe.

YOUNG RUGELEY, (1325), l. in 1880; br. W. Lander, s a Beach Ram, d by Sir Roger (1109).

YOUNG SALOPIAN, (1326), l. in 1880; br. T. J. Mansell, s Pride of Montford (959), d by County Member (452).

YOUNG SECUNDUS, (1327), l. in 1875; br. Viscount Falmouth s Secundus (1049), d a Mansell Ewe.

YOUNG VICTOR, (1329), l. in 1880; br. S. C. Pilgrim, s Victor (150), d by Ambition (Pilgrim).

YOUNG VOLUNTEER, (1330), l. in 1868; br. E. Thornton, s Volunteer (51), d by Canterbury Patentee 2nd (S. Byrd).

YOUNG YATES, (1331), l. in 1873; br. Viscount Falmouth, s Harry (656).

LIST OF SIRES
USED IN THE FLOCKS OF THE FOLLOWING BREEDERS
STATING YEAR OF SERVICE.

—◆—

Mr. JOSEPH ROBINSON ADDERLEY, Teddesley Coppice, Penkridge.

1877—Captain (332)	1880—Traveller (1233)
1878—Captain (332)	1881—Big Back (253)
Trickster (1236)	Traveller (1233)
1879—Captain (332)	1882—Big Back (253)
Traveller (1233)	Hardy (650)
Trickster (1236)	Traveller (1233)

Mr. GEORGE ALLEN, Yew Tree Farm, Penkridge.

1878—Aristocrat (192)	1881—Graphic (631)
1879—Aristocrat (192)	1882—Banker (208)
1880—Aristocrat (192)	Challenge (347)
Graphic (631)	Improver (678)
1881—Brewood (272)	

Sir HENRY ALLSOPP, Bart., Hindlip Court, Worcester.

1880—Baron Pendeford (217)	1882—Chesham's No. 8, 1882
1881—Baron Pendeford (217)	(376)
Beach's No. 12, 1881	Chesham's No. 14, 1882
(232)	(377)
1882—Baron Pendeford (217)	Loder's No. 11, 1882 (724)
Beach's No. 2, 1882 (233)	

Mr. JOHN H. ARKWRIGHT, Hampton Court, Leominster.
1882—Bach's No. 7, 1882 (200)

Mr. FRANCIS BACH, Onibury, Craven Arms.

1869—Paddy (929)	1874—Standard (1158)
1870—Paddy (929)	The Warden (1210)
1871—Paddy (929)	1875—Oswestry Champion (137)
The Warden (1210)	The Warden (1210)
1872—Paddy (929)	1876—Oswestry Champion (137)
The Warden (1210)	Tartar (1173)
1873—Paddy (929)	1877—Oswestry Champion (137)
Standard (1158)	Sir Joseph (1101)
The Warden (1210)	Tartar (1173)

1878—Oswestry Champion (137)
    Sir Joseph (1101)
1879—Cockney (401)
    Lord Kilburn (149)
    Oswestry Champion (137)
    Yeoman (1287)
1880—Blood Royal (261)
    Cockney (401)
    Oswestry Champion (137)
    Yeoman (1287)

1881—Cockney (401)
    Major (793)
    Oswestry Champion (137)
    Yeoman (1287)
1882—Cockney (401)
    Double R. (172)
    Major (793)
    Maximus (836)
    Yeoman (1287)

MR. WILLIAM BAKER, Moor Barns, Atherstone.

1869—Hero (667)
1870—Hero (667)
1871—Hero (667)
1877—Goliah (611)

1881—Royal Reserve (159)
1882—Caldicote (321)
    Harlescote 2nd (654)
    Royal Reserve (159)

MR. RICHARD BARBER, Harlescott, Shrewsbury.

1866—Vulcan (1262)
1868—Shrawardine (1077)
1870—Cronkhill (482)
1871—Uffington (1250)
1872—Latimer (701)

1874—Son of Cardinal (1126)
1877—Cardinal 2nd (340)
    Lord Claud (741)
1878—Montford (870)
1881—Richmond (1005)

MR. JOHN ADCOCK BARRS, Nailstone, Hinckley.

1875—Odstone (918)
1876—Lord Beaconsfield (736)
    Odstone (918)
1877—Count (449)
    Corporal (443)
    Lord Beaconsfield (736)
1878—Corporal (443)
    Czar (488)
    Freeford (582)
    Lord Beaconsfield (736)
1879—Berkeley (248)
    Czar (488)
    Marquis of Lorne (157)
    The Ruler (1205)

1880—Berkeley (248)
    Czar (488)
    Lord Oxon (167)
    Marquis of Lorne (157)
1881—Berkeley (248)
    Czar (488)
    Marquis of Lorne (157)
    Pride of Freeford (957)
1882—Beaudesert (239)
    Berkeley (248)
    Chieftain (385)
    Cromwell (481)
    Czar (488)
    Dignity (498)

Mrs. MARIA BARRS, Odstone Hall, Atherstone.

1873—Major (791)
1874—Magnet (787)
1875—Odstone (918)
1876—Lord Beaconsfield (736)
Odstone (918)
1877—Corporal (443)
Count (449)
Lord Beaconsfield (736)
Lord Odstone (766)
1878—Corporal (443)
Czar (488)
Freeford (582)
Lord Beaconsfield (736)
1879—Berkeley (248)

1879—Czar (488)
The Ruler (1205)
1880—Berkeley (248)
Czar (488)
Lord Oxon (167)
1881—Berkeley (248)
Czar (488)
Pride of Freeford (957)
1882—Beaudesert (239)
Berkeley (248)
Candidate (325)
Cromwell (481)
Czar (488)
Dignity (498)

Mr. CHAS. H. BASSETT, Pilton House, Barnstaple.

1881—Yardley, No. 10 (1284)
1882—Montford, No. 8 (872)

1882—Yardley, No. 10 (1284)

Mr. JOSEPH BEACH.
(See the late Mr. Joseph Beach).

The late Mr. JOSEPH BEACH, The Hattons, Brewood, Stafford.

1865—Gayton Prince (592)
1866—Gayton Prince (592)
Young Gayton (1307)

1867—Cardinal (53)
Gayton Prince (592)
Young Gayton (1307)

At this period the above Flock in its entirety came into the possession of

Mrs. SARAH BEACH, The Hattons, Brewood.

1868—Briton (296)
1869—Briton (296)
Duke of Manchester (70)
1870—Briton (296)
Duke of Manchester (70)
Quickstep (987)
The Knight (1190)
Wrinkle Face (1282)
1871—Major (88)
1872—Briton (296)
Lord Stafford (775)
Sir Roger (1109)
1873—Cardigan (338)

1873—Lord Stafford (775)
Lord Taunton (121)
Reflection (998)
1874—British Oak (290)
Cardigan (338)
1875—Masterman (828)
Monarch (863)
1876—British Oak (290)
Colossus (411)
Manager (795)
Masterman (828)
Monarch (863)

At this period the above Flock came into the possession of
MR. JOSEPH BEACH, The Hattons, Brewood.

1877—Captain Thomas (333)
    Challenger (348)
    Sir Garnet (1091)
    Snowflake (1120)
1878—Sheldon's No. 2 (143)
1879—First Choice (561)
    Seaman (1047)
1880—Lord Falmouth (746)

1880—Prince Imperial (966)
    Royal Chief (1022)
1881—Challenge (347)
    Minton's Pride (858)
    Royal Chief (1022)
1882—Pride of Brewood (955)
    Pride of Hatton's (958)
    Royal Chief (1022)

MR. GEORGE JOHN BELL, The Nook, Irthington, Carlisle.

1878—Robin Rough Nob (1017)
1879—Lord Kineton (752)
    Robin Rough Nob (1017)
1880—Lord Kineton (752)
1881—King of the Heath (697)

1881—Lord of the Heath (768)
1882—Cock of the North (402)
    King of the Heath (697)
    Lord of the Heath (768)

MR. HUGH BENNETT, Builth Wells, Breconshire.

1882—Newry (893) | 1882—Prince Lewellin (968)

MR. A. S. BERRY.—*(See Mr. T. Ryland).*

MR. JOHN BOURNE, The Arbour Farm, Market Drayton.

1877—Brigand (277)
1880—Hawkins, No. 6, 1880 (658)

1881—Hawkins, No. 6, 1880 (658)
1882—Brontes (301)

MR. J. BOWEN-JONES.—*(See Mr. Evan Bowen).*

MR. EVAN BOWEN, Ensdon House, Shrewsbury.

1860—Chester Billy (7)
1861—Chester Billy (7)

1862—Chester Billy (7)

At this period the above Flock in its entirety came into the possession of

MESSRS. BOWEN & JONES, Ensdon House, Shrewsbury.

1863—Chester Billy (7)
1864—Patentee the Prime (28)
1865—Patentee the Prime (28)
1866—Patentee the Prime (28)
1867—Lord Warden (777)
    Worcester Patron 2nd (1281)
1868—Lord Warden (777)
    Mountain 2nd (875)
    Worcester Patron 2nd (1281)

1869—Son of Lord Clifden (1141)
    Worcester Patron 2nd (1281)
1870—Fosse Duke (576)
    Mountain 3rd (876)
    Son of Lord Clifden (1141)
1871—Mountain 3rd (876)
1872—Ensdon Hero (104)
    Foreman (568)
    Mountain 3rd (876)

At this period the above Flock in its entirety came into the possession of

MR. J. BOWEN-JONES, Ensdon House, Shrewsbury.

1873—Conservative Calcot (436)
    Evans' No. 3, 1873 (534)
1874—Aristocrat (191)
    Conservative Calcot (436)
    Son of Little Lord (1139)
1875—Aristocrat (191)
    Calcot 2nd (318)
    Claudius (103)
1876—Claudius (103)
    Crane's No. 6, 1876 (470)
    Mansell's No. 16, 1876 (802)
1877—Bedford Hero (111)
    Claudius 2nd (395)
    Mansell's No. 16, 1876 (802)
1878—Claudius 2nd (395)
    Dudmaston No. 21, 1878 (507)
    Mansell's No. 16, 1876 (802)

1878—Severus 2nd (1052)
1879—Dudmaston No. 21, 1878 (507)
1879—Mansell's No. 16, 1876 (802)
    Royal Reserve (159)
1880—Dudmaston No. 21, 1878 (507)
    Duke Goliah (508)
    Ercall P. (528)
    Leviathan 4th (713)
    Marquis of Bath (822)
1881—G.C.B. (593)
    K.C.B. (693)
    Marquis of Bath (822)
    Salopian C (1039)
    Treble C (1234)
1882—G.C.B. (593)
    K.C.B. (693)
    Salopian C (1039)
    The Patriot Lord (1199)

MESSRS. BOYD, Bloomfield, Co. Down Ireland.

1881—Wellington 4th (1271) | 1882—Wellington 4th (1271)

Mr. RICHARD BROMLEY, Felton Butler, Montford Bridge, R.S.O., Shropshire.

1872—Hardlines (647)	1880—Thomas No. 17, 1880
1873—Hardlines (647)	(1215)
1874—Hardlines (647)	Wonder (1277)
1875—Hardlines (647)	Hardy (649)
1876—Challenger (348)	1881—Hardy (649)
1877—Lord Harlech (135)	Harold 2nd (655)
1878—Lord Harlech (135)	Thomas No. 17, 1880
Wonder (1277)	(1215)
1879—Lord Harlech (135)	1882—Hardy (649)

Mr. JOSEPH BROWN, Bednall, Stafford.

1879—Big Gun (255)	1882—Baron (212)
1880—Baron (212)	Bednall (241)
Bednall (241)	Bednall Hero (242)
1881—Baron (212)	Bread Winner (269)
Bednall (241)	

Mr. RICHARD BROWN, Ruyton-xi-Towns, Shropshire.

1880—Bentick (247)	1882—Active (178)
1881—Bentick (247)	Prince Victor (158)

Mr. JOHN G. BUCHANAN, Coston Hall, Aston-on-Clun, Shropshire.

1880—A1 (176)	1881—Banjo (206)
Coston (447)	1882—Chatsworth (360)

Mr. CHARLES BYRD, Littywood, Stafford.

1865—Canterbury 2nd (328)	1869—Blood Royal (83)
Canterbury 3rd (329)	Young Guardian (1309)
Littywood Patentee (721)	Young Quality (1322)
Monarch (866)	1870—Commander (415)
The Norman (1197)	Major (88)
1866—Canterbury 3rd (329)	Manager (794)
Lord Weston (778)	Young Napier (1319)
Monarch (866)	1871—Legatee (95)
1867—Canterbury 3rd (329)	Oxford Hero (77)
Guardian (642)	1872—Defiance (496)
Viceroy (66)	Ensign (527)
1868—Guardian (642)	Little Lord (718)
Lord Grey (749)	Lord Knightley (753)
Lord Napier (764)	Tory (1227)

1873—Apollo (190)
    Earl of Chester (519)
    The General (1186)
    Tory (1227)
1874—Captor (334)
    Earl of Chester (519)
    Hampton Hero (84)
1874—Young General (1308)
1875—Captor (334)
    Chivalry (387)
    Young General (1308)
1876—Lord Latimer (754)
    Touchstone (1230)
    Viscount (1259)
1877—Birmingham Reserve(126)
    Challenger (349)
    Touchstone (1230)

1878—The Hero (1188)
    Victor (150)
1879—Beaufort (240)
    The Hero (1188)
    True Blue (1239)
    Typical (1248)
1880—Beaufort (240)
    Cossack (445)
    Hermit (666)
    The Hero (1188)
    Thickset (1213)
1881—Chancellor (355)
    Hermit (666)
    Thickset (1213)
    Truine (1238)
1882—Beach's No. 3, 1882 (234)
    Sterling (164)

MR. D. CAMERON, Fettes, by Inverness.
1882—Number Six (917)

MR. THOS. CARTWRIGHT, The Buildings, West Felton,
R.S.O., Shropshire.

1871—Cicero (389)
1872—Cicero (389)
1874—Big Rednal (256)
    Ellesmere (523)
1875—Big Rednal (256)
    Evans' Ram (535)
1876—Minton's Ram (860)
1878—Celerity (345)
    Masterman (829)

1879—Celerity (345)
    Prince Knightley, (967)
1880—Celerity (345)
    Prince Knightley (967)
1881—Birkenhead (257)
    Prince Knightley (967)
1882—Birkenhead (257)
    The Victor (1209)

THE RIGHT HON. THE EARL CAWDOR, Stackpole Home Farm,
near Pembroke, S. Wales.

1880—Sir Richard (1106)
    Sir William (1116)
1881—Sam (1041)
    Sir Joseph (1102)

1882—Flag of Britain (565)
    Sam (1041)
    Sir Joseph (1102)
    Sir Robert (1105)

MR. T. F. CHEATLE, Dosthill, Tamworth.

1867—Iron Duke (681)
1868—Iron Duke (681)
    Pirate (938)
1869—Pirate (938)
1870—Reindeer (1001)
1871—Reindeer (1001)
1872—Confederate (429)
    Reindeer (1001)
1873—Confederate (429)
1874—Confederate (429)
1875—Germans No. 6, 1875
    (603)
1876—Germans No. 6, 1875
    (603)

1877—Germans No. 6, 1875
    (603)
1878—Beaconsfield (236)
    Son of Hero (1136)
1879—Beaconsfield (236)
    Son of Hero (1136)]
1880—Beaconsfield (236)
    Warrior (1265)
    Young Clinker (1298)
1881—Anak (187)
    Young Clinker (1298)
1882—Anak (187)
    Conqueror (434)

LORD CHESHAM, Latimer, Chesham, Bucks.

1880—Cardinal (339)
    Champion (352)
    Chancellor (354)
    Colonel (408)
    Comus (426)
    Dudmaston (506)

1881—Banker (207)
    Bannerman (209)
    Baron (213)
    Benedict (244)
    Royal Grand Duke (1025)
1882—Cavalier (344)

MR. C. F. CLARK, Perton Grove, near Wolverhampton.

1872—The Knight (1191)
1873—True Born (1240)
1874—Young Ensdon (1306)
1875—German (602)
1876—Minton's No. 6, 1876
    (845)
1877—Yates' No. 13, 1877
    (1286)

1878—Coxon's No. 14, 1878
    (461)
1879—His Lordship 2nd (670)
1880—Lord Leicester (755)
1881—Magnum Bonum (790)
1882—Coxswain (403)
    Lord de Lisle (744)

MR. M. W. COLCHESTER-WEMYSS, The Wilderness, Mitcheldean, Gloucestershire.

1881—Colonel (409)
    General (597)

1882—Colonel (409)
    General (597)

Mr. EDWARD MOUNTFORD COLEMAN, Oakenshaw,
Headless Cross, Redditch.

1877—Noble (897)
1878—Lord of the Isles (770)
1879—Cyprus (486)
1880—Shrewsbury Hero (1082)
1881—Lord Rugby (774)
 Mansell's No. 12, 1881 (811)

1881—Master May (833)
 Minton's No. 6, 1881 (850)
1882—Ipsley's Pride (680)
 Lord Oakenshaw (765)
 North Star (913)
 Redditch Wonder (996)
 Studley Hero (1165)

Mr. GRIMWOOD COOKE, Horseheath Park, Cambs.

1882—Firmament (560)
 Floss (567)
 Foundation (580)

1882—Frere (586)
 Fringe (588)
 Full Eye (589)

Mr. WILLIAM COTTERELL, Derry Ormond Park,
Cardiganshire.

1880—Benedict (245)
1881—Benedict (245)
 Master Dick (827)

1882—Benedict (245)
 Dandy (490)
 Master Dick (827)

Mr. CHARLES COXON, Elford Park, Tamworth.

1877—Formation (572)
 Foundation (579)
1878—Debtor (495)
1879—Popularity (941)
 Regulator (1000)
1880—Popularity (941)
 Regulator (1000)
 Young Masterpiece (1317)
1881—Beaudesert (239)

1881—Compass (424)
 Economy (522)
 Popularity (941)
 Young Masterpiece (1317)
 Young Victor (1329)
1882—Compass (424)
 Economy (522)
 Hero (668)

Mr. JOHN COXON, Freeford, Lichfield.

1857—Veteran (1255)
1858—Veteran (1255)
1859—Lord Flash (747)
 Patent (931)
 Valiant (1254)
 Veteran (1255)
1860—Valiant (1254)
1864—Duke of Newcastle (39)
1866—Sheet Anchor (1060)
1867—Clipstone (399)

1867—Commander (415)
1868—Commander (415)
 Mansion 3rd (818)
1869—Commander-in-Chief (416)
1870—Apollo (190)
 Commander-in-chief (416)
1871—Apollo (190)
 Commander-in-chief (416)
1872—Stamina (1157)
1874—Capitalist (331)

1874—Champion (351)
1875—Capitalist (331)
Champion (351)
Colossus (411)
1876—Magnet (788)
1877—First Fruits (564)

1878—Masterpiece (832)
Sir Robert (1108)
1879—City Member (391)
1880—Cockade (400)
1881—Countryman (451)

Mr. JOHN LAWRENCE COXON, Tempe House, Shepstone, Ashby-de-la-Zouch

1872—Utility (1253)
1873—Premier (949)
1874—Fat Back (552)
1875—Cossack (446)
1876—Leviathan (710)
1877—Bouncer (264)
1878—Courtier (455)

1879—Sterling (1162)
1880—Cardiff (337)
1881—Cardiff (337)
Earldom (517)
1882—Mountain (874)
Waterproof (1269)

Messrs. CRANE & TANNER.—*(See Messrs. J. & E. Crane).*

Messrs. J. & E. CRANE, Shrawardine, Montford Bridge, R.S.O., Shropshire.

1854—Tern (1176)
1855—Tern (1176)
1856—Tern (1176)
1857—Patentee (4)
1858—Celebrity (6)
1860—Caradoc (336)
1861—Caradoc (336)
1862—Caradoc (336)
1863—Chieftain (384)
Nobleman (37)
1864—Chieftain (384)
Nobleman (37)
Pride of Pitchford (961)
1865—Chieftain (384)
Nobleman (37)
Plymouth Prize (47)
1866—Chieftain (384)
Big Plymouth (49)

1866—Plymouth Prize (47)
Shamrock (1054)
Sheet Anchor (1060)
1867—Big Plymouth (49)
Caractacus (335)
Duke of Newcastle (39)
Shamrock (1054)
1868—Caractacus (335)
Corsair (57)
Duke of Newcastle (39)
Lord Uffington (56)
1869—Caractacus (335)
Crosswood Hero (65)
Pretender (951)
1870—Royalty (1029)
Young Caractacus (1295)
1871—Young Caractacus (1295)

At this period the above Flock in its entirety came into the possession of

Mr. EDWARD CRANE, Shrawardine, Montford Bridge, R.S.O., Shropshire.

1871—Cambrian (324)	1874—Claudius (103)
1872—Claudius (103)	Landseer (700)
1873—Caligula (112)	1875—Caligula (112)
Claudius (103)	Claude Duval (394)
1874—Chivalry (387)	1876—Caligula (112)

At this period the above Flock in its entirety came into the possession of

Messrs. CRANE & TANNER, Shrawardine, Montford Bridge, R.S.O., Shropshire.

1876—Son of Chivalry (1128)	1879—Yardley (1283)
1877—Columbus (128)	1880—Bristol Prince (285)
Coxcomb (457)	Dudmaston (504)
Dudmaston (504)	1881—Dudmaston (504)
1878—Bristol Reserve (144)	Montford Hero (160)
Columbus (128)	Simon de Montford (1089)
Dudmaston (504)	1882—Baron Plassy (218)
1879—Columbus (128)	Dudmaston (504)
Dudmaston (504)	Minton's Pride (858)

Mr. H. CRAWSHAY, Stormer Hall, Leintwardine, Herefordshire.

1879—Stormer (1163)	1881—Brinks (279)
1880—Brinks (279)	Stormer (1163)
Stormer (1163)	1882—Brinks (279)

Mr. WILLIAM T. CRAWSHAY, Cyfarthfa Castle, Merthyr Tydfil, S. Wales.
1882—Sir Garnet (1092)

Major R. W. CRADOCK, Derrycalaghan, Roscrea, Ireland.
1882—Police Constable (940)

Mr. WILLIAM CURETON, The Beam House, Montford Bridge, R.S.O., Shropshire.

1881—Harding's No. 10, 1881 (687)	1882—Crane and Tanner's No. 7, 1882 (480)
Lee's No. 1, 1880 (705)	Harding's No. 10, 1881 (687)

Mr. JOHN DARLING, Beau Desert, Rugeley, Staffordshire.

1878—Grand Turk (629)
    Monarch (864)
1879—Grand Turk (629)
    Monarch (864)
1880—Dudmaston Hero (165)

1881—Bristol Duke (283)
    Cockade (400)
1882—Lord Coxcomb (743)
    Mansell's No. 3, 1882
    (812)

Mr. EDWARD DAVIES, Gatacre Park, Bridgnorth, Shropshire

1881—Young Rugeley (1325)
1882—Chadbury (346)

1882– Young Rugeley (1325)

Mr. WILLIAM M. DAWES, New House, Craven Arms, Shropshire.

1879—Oney (922)
1880—Oney (922)

1882—Blackall (259)
    Brisbane (280)

Mr. HAROLD DEWHURST, Aireville, Skipton, Yorkshire.

1882—Brabazon (266)

1882—Lavender (704)

Mr. THOMAS DICKEN, Ellerdine, Wellington, Shropshire.

1875—Byrd's No. 18, 1875 (314)
    Evan's No, 12, 1875 (536)
1876—Beach's No. 15, 1876 (227)
    Truestock (1243)
1877—Beach's No. 15, 1876 (227)
    Count (450)
    Mansell's No. 10, 1877
    (804)
1878—Count (450)
    Mansell's No. 10, 1877
    (804)
    Mansell's No. 22, 1878
    (807)
1879—Crane and Tanner's No.
    8, 1879 (474)
    Mansell's No. 10, 1877
    (804)
    Mansell's No. 22, 1878
    (807)

1879—Sheldon's Ram (1070)
1880—Crane and Tanner's No.
    8, 1879 (474)
    Lord Kilburn (149)
    Mansell's No. 22, 1878
    (807)
1881—Allsopp's No. 2, 1881
    (184)
    Condor (428)
    Minton's No. 20, 1881
    (853)
    Quality (984)
1882—Condor (428)
    Mansell's No. 7, 1882
    (813)
    Pulley's No. 1, 1881
    (980)

Mr. WILLIAM CHARLES DOUTHWAITE, Rousham, Banbury, Oxon.
1882—Oxford Hero (926)

MR. GEORGE EDWARDS, Little Brampton, via Kington, Herefordshire.

1879—Sir Roger (1111)
1880—Sir Roger (1111)
Wonderful (1278)
1881—Sir Roger (1111)

1881—Wonderful (1278)
1882—Royal Ensign (1023)
Sir Joseph (1103)

MR. GEORGE WITHERS EDWARDS, JUNR., Woolston, Oswestry, Shropshire.

1879—Earl of Shrewsbury (521)
Thomas' No. 2, 1879 (1218)
1880—Earl of Shrewsbury (521)
Thomas' No. 2, 1879 (1218)
Thomas' No. 19, 1880 (1220)

1881—Earl of Shrewsbury (521)
Minton's No. 14, 1881 (852)
1882—Earl of Shrewsbury (521)
Minton's No. 10, 1882 (856)

MR. JOHN W. EDWARDS, The Court, West Felton, R.S.O., Shropshire.

1873—Defiance (496)
1874—Defiance (496)
1875—Defiance (496)
1876—Cormorant (442)
1877—Cormorant (442)
1878—Cormorant (442)
1879—Thomas' No. 2, 1879 (1218)

1880—Thomas' No. 2, 1879 (1218)
Thomas' No. 19, 1880 (1220)
1881—Minton's No. 14, 1881 (852)
1882—Earl of Shrewsbury (521)
Minton's No. 10, 1882 (856)

MR. JOHN EVANS, Uffington, Shrewsbury.

1851—Own Brother to Bossy (925)
1852—Own Brother to Bossy (925)
1853—Own Brother to Bossy (925)
1854—Own Brother to Bossy (925)
1855—Emperor (525)
Own Brother to Bossy (925)
1856—Emperor (525)
1857—Emperor (525)
1858—Emperor (525)

1858—Humphrey Davy (676)
1859—Emperor (525)
Humphrey Davy (676)
1860—Emperor (525)
Humphrey Davy (676)
Young Emperor (1305)
1861—Humphrey Davy (676)
Sir Samuel (1113)
Young Emperor (1305)
1862—Sir Samuel (1113)
Young Emperor (1305)
1863—Pride of Pitchford (961)
Sir Samuel (1113)
Young Emperor (1305)

1864—Competition (425)
Sir Samuel (1113)
Young Emperor (1305)
1865—Competition (425)
Nonpareil (908)
Pride of Pitchford (961)
Volunteer (51)
Young Emperor (1305)
1866—Competition (425)
Emigrant (524)
Nonpareil (908)
1867—Competition (425)
Emigrant (524)
Nonpareil (908)
1868—Chieftain (384)
Competition (425)
Nonpareil (908)
Premier (947)
1869—Abbot of Bury (54)
Cardinal (53)
Chieftain (384)
Favourite (553)
Lord Uffington (56)
Premier (947)
Standard Bearer (80)
1870—Cardinal (53)
Hardlines (647)
Lord Uffington (56)
Royalty (1029)
1871—Broadgauge (89)
Cardinal (53)
Grand Duke (620)
Rob Roy (1018)
1872—Broadgauge (89)
Grand Duke (620)
Hardlines (647)
Proud Salopian (87)
Union Jack (1252)
Young Cambrian (1294)
1873—Broadgauge (89)
Claudius (103)

1873—Grand Duke (620)
Hardlines (647)
Proud Salopian (87)
Union Jack (1252)
Young Cambrian (1294)
1874—Cavalier (343)
Grand Duke (620)
Paddy Green (930)
Union Jack (1252)
1875—British Oak (290)
Grand Duke (620)
Paddy Green (930)
Union Jack (1252)
1876—British Tar (292)
Cavalier (343)
Grand Duke (620)
Truelight (1242)
1877—British Oak (290)
May Duke (837)
Royal Taunton (115)
1878—British Oak (290)
May Duke (837)
Royal Taunton (115)
1879—Bacchus (197)
Bristol Reserve (144)
May Duke (837)
Royal Taunton (115)
1880—Baron Bristol (215)
Bristol Reserve (144)
Lord Coxcomb (743)
May Duke (837)
Royal Taunton (115)
1881—Baron Bristol (215)
Lord Coxcomb (743)
Royal Gem (1024)
Royal Taunton (115)
1182—Bristol Chieftain (282)
Bristol Reserve (144)
Grand Chief (619)
Lord Coxcomb (743)

MR. PETER EVERALL, Uckington, Shrewsbury.

1874—Minton's No. 1, 1874 (844)
1875—Minton's No. 1, 1874 (844)
1876—Minton's No. 1, 1874 (844)
1877—Brigadier (275)

1877—Minton's No. 1, 1874 (844)
1878—Cardinal 2nd (340)
1879—Prince Victor (158)
1880—Hercules (663)
Lord Carlisle (153)
1881—Crane & Tanner's No. 5, 1881 (477)
Lord Carlisle (153)

1881—Sterling (164)
1882—Chesham's No. 2, 1882 (374)
Chesham's No. 5, 1882 (375)
Second Best (170)
Son of Lord Carlisle (1140)

Mr. M. G. FARRER, The Park Farm, Great Malvern.

1880—Friar Tuck (587)
1881—Friar Tuck (587)
Robin Hood (1014)

1882—Little John (716)
Robin Hood (1014)

The Right Hon. VISCOUNT FALMOUTH, Tregothan, Probus, Cornwall.

1858—Chester (381)
1860—Young Chester (1296)
1862—Little Chester (715)
1870—Lord Paramount (772)
Nock's No. 3, 1870 (903)
Nock's No. 4, 1870 (904)
Tamworth (1172)
Yates No. 6, 1870 (1285)
1872—Harry (656)
Jemmy (686)
Sam (1040)
1873—Bachelor (199)
Sheriff (1073)
1874—Freeford (583)
Gipsy King (607)
Lichfield (714)
Luck's All (782)
Nobleman (900)
Secundus (1049)
Sir Robert (1107)
Sir Roger (1110)
Turk (1246)
Young Yates (1331)

1876—King Harry (696)
Robertson (1012)
Young Secundus (1327)
1877—Coldstream (406)
General (594)
Masterman (828)
The Colonel (1179)
1878—Gladiator (608)
Hatton's British Oak (657)
Orangemore (923)
Young Bachelor (1289)
1879—Goliath (612)
Masterpiece (831)
1880—British Tar (293)
Grandmaster (624)
Yardley (145)
1881—Beach's No. 10, 1881 (231)
Charles 2nd (359)
Comet (414)
Heart of Oak (660)
Lord Falmouth (746)
1882—Pocket Hercules (939)

The following Rams have been used in the above Flock, but the dates of service are not given.

Happy-go-Lucky (645)
Long Sheep (730)

Lord Mayor (758)
Samson (1043)

Mr. JOHN EDWARD FARMER, Felton, Ludlow, Shropshire.

1873—Beach's No. 3, 1873 (221)
1876—Leinster (706)
    Lord Acton (731)
    Randell's No. 1, 1876 (988)
1877—Speculator (1153)
1878—Double X (501)
    Heart of Oak (659)
    Priam (952)
    Wigley Duke (1272)
1879—Fenn's No. 4, 1879 (556)
    Goliah 2nd (613)
1880—Carlisle (166)
1881—Carlisle (166)

1881— Lusty (783)
    Mansell's No. 11, 1881 (810)
    Marquis of Montford (823)
    May Fly (838)
    Reliance (1002)
    Safety Valve (1037)
    Warrior (1267)
1882—Carlisle (166)
    Coxon's No. 20, 1882 (463)
    Crane & Tanner's No. 6, 1882 (479)

Mr. J. W. FAUX, Coleorton, Ashby-de-la-Zouch.

1881—The Prior (1202)
1882—Omar Pasha (921)

1882—The Prior (1202)

Mr. THOMAS FENN, Stonebrook House, Ludlow, Shropshire.

1868—Abbot of Bury (54)
    Kenyon (695)
1869—Novelty (41)
    Downton Pippin (502)
1870—Kingscraft (698)
1871—Marquis (820)
1872—Marquis (820)
    Midlothian (842)
1873—Bruce (304)
    Marquis (820)
    Midlothian (842)
1874—Ensdon Hero (104)
    Midlothian (842)
1875—Ensdon Hero (104)
1876—Birmingham Reserve (126)
    Ensdon Hero (104)
1877—Beach's No. 17, 1876 (228)
    Ensdon Hero (104)
    Son of Claudius (1130)
1878—Lord Aston (123)
    Lord Beaconsfield (737)
    Bacchus (197)

1879—Lord Aston (123)
    Lord Beaconsfield (737)
    Sheldon's No. 4, 1879 (1065)
1880—Caligula 2nd (323)
    Clinker (397)
    Lord Odstone (766)
    Sir Garnet (1092)
1881—Bacchus 2nd (198)
    Clinker (397)
    Lord Oxon (167)
    Montford Hero (160)
    Sir Garnet (1092)
1882—Bacchus 2nd (198)
    Cockney (401)
    Duke Goliah (508)
    Lord Oxon (167)
    Mansell's No. 8, 1882 (814)
    Sir Evelyn (1090)
    Symmetry (1171)
    The Patriot Lord (1199)

Mr. WILLIAM FOWLER, Acton Reynald, Shrewsbury.

1876—Chesham's No. 19, 1876
(368)
1877—Chesham's No. 19, 1876
(368)
Mansell's No. 1, 1877,
(*Smithfield Sale*) (803)
1878—Chesham's No. 19, 1876
(368)
1879—Crane and Tanner's No.
2, 1879 (*Smithfield
Sale*) (473)
Wonder (1277)
1880—Chandos (356)

1880—Crane and Tanner's No.
2, 1879 (*Smithfield
Sale*) (473)
Oxon Caligula (927)
1881—Chandos (356)
Oxon Caligula (927)
Thomas' No. 4, 1881
(1222)
1882—Sir Gregory (1097)
The Earl's No. 14, 1882
(1185)
Thomas' No. 4, 1881
(1222)

Mr. THOMAS FRANK, Cound Arbour, Shrewsbury.

1882—Instone's Ram (679) | 1882—Shropshire Ram (1087)

The Executors of the late Mr. WILLIAM GERMAN.

1866—Novelty (41)
Sweet William (1170)
1867—The Peer (1200)
Young Duke, (1303)
1868—Lord Lincoln (756)
Preserver (950)
1869—Chancellor (353)
Multum-in-Parvo (884)
Pensioner (935)
Young Conservative (1302)
Young Volunteer (1330)
1870—Captivator (76)
Cohesion (405)
Commodore (418)
Conductor (90)
1871—Apollo (190)
Rifleman (1008)
Safeguard (1035)
1872—Crown Prince (483)
Lord Haughton (751)
Stamina (1157)
1873—Confidence (101)
Consolation (438)
Neptune (890)

1873—Young Claudius (1297)
1874—Claimant (392)
Crane's No. 11, 1873 (468)
Ranger (992)
1875—Champion (351)
Colossus (411)
Conserver (437)
1876—Courtier (454)
Leviathan (709)
Magnet (788)
1877—First Fruits (564)
Marquis of Bath (822)
Sir Robert (1108)
The Czar (1182)
1878—Farmer's Friend (548)
Lord Mona (762)
Masterpiece (832)
Son of Champion (1127)
1879—City Member (391)
Crane and Tanner's No.
18, 1879 (476)
Lord of the Manor (771)
Lord Oxon (167)
Quality (983)

1879—Speculation (1152)
1881—Brilliant (278)
    Corporal (443)
    Coxon's No. 4, 1880 (462)
    German's No, 16, 1881 (604)
    Macintosh (786)
    Marquis of Lorne (157)

1881—Masterman (830)
    Pride of Freeford (957)
    Robin Rough (1016)
1882—Brewood Chief (273)
    Challenger (350)
    Ram Lamb (991)
    Robin Rough (1016)

Mr. ARTHUR S. GIBSON, Bulwell, Notts.

1880—Chesham (361)
    Gallant (590)
1881—Gallant (590)

1881—Shrawardine (1079)
1882—Gambetta (591)
    Gold Dust (610)

Mr. FRANCIS GIBSON, Woolmet, near Dalkeith, N.B.

1881—Mutineer (886)
1882—Jock (688)
    Mutineer (886)

1882—Shropshire Ram (1088)
    Sulphur (1166)

Messrs. J. B. & G. H. GREEN.—*see Mr. J. B. Green.*

Mr. J. B. GREEN, Marlow Lodge, Leintwardine, Herefordshire.

1868—Crane's No. 11, 1868 (465)
    Crane's No. 17, 1868 (466)
1869—Crane's No. 11, 1868 (465)
    Crane's No. 17, 1868 (466)
1870—Crane's No. 11, 1868 (465)
1872—Evans No. 1, 1872 (532)
    Evans No. 5, 1872 (533)

1873—Evan's No. 1, 1872 (532)
    Evans No. 5, 1872 (533)
    Nightingale's No. 1, 1873 (894)
1874—Evans No. 5, 1872 (533)
    Mansell's No. 21, 1874 (799)
1875—Nightingale's No. 1, 1873 (894)
1876—Mansell's No. 21, 1874 (799)

At this period the Flock in its entirety came in the possession of

Messrs. J. B. & G. H. GREEN, Marlow Lodge, Leintwardine, Herefordshire.

1877—Nightingale's No. 1, 1873 (894)
1878—Farmer's No. 1, 1878 (542)

1878—Farmer's No. 2, 1878 (543)
1879—Farmer's No. 1, 1878 (542)

1879—Farmer's No. 2, 1878 (543)
Green's No. 22, (632)
1880—Farmer's No. 2, 1878 (543)
Fenn's No. 16, 1880 (557)
Little Nightingale (720)
1881—Farmer's No. 1, 1878 (542)
Farmer's No. 36, 1881 (547)
Green's No. 22, (632)
Green's No. 21, 1879 (633)
Green's No. 12, 1880 (635)
Green's No. 6, 1881 (636)

1881—Green's No. 12, 1881 (637)
Minton's No. 24, 1881 (854)
1882—Bach's No. 18, 1882 (201)
Farmer's No. 1, 1878 (542)
Fenn's No. 29. 1882 (558)
Green's No. 22 (632)
Green's No. 18, 1881 (634)
Green's No. 10, 1882 (638)
Green's No. 12, 1882 (639)
Green's No. 18, 1882 (640)
Little Nightingale (720)

MR. THOMAS GREGORY, Eyam View, Eyam, Derbyshire.
1882—Clansman (393)

MRS. MARY GROVES,—*see the late Mr. Robert Groves.*

The late MR. ROBERT GROVES, Berrington, Shrewsbury.

1870—Volunteer 2nd (1261)
1871—Volunteer 2nd (1261)
1872—Volunteer 2nd (1261)
1873—Berrington Hero (249)
1874—Berrington Hero (249)
1875—Berrington Hero (249)
1876—Betton (252)
Grand Master (623)

1877—Betton (252)
Grand Master (623)
1878—Betton (252)
Grand Master (623)
1879—M.P. (879)
1880—Mark Anthony (819)
M.P- (879)
Thomas' No 1, 1879 (1217)

At this period the Flock in its entirety came into the possession of
MRS. MARY GROVES, Berrington, Shrewsbury.

1881—Mark Anthony (819)
Thomas' No. 1, 1879 (1217)

1882—Baker Pacha (205)
Rifleman (1009)

MR. JOSEPH HALL, Birchdale Farm, Frodsham, Cheshire.
1882—Lordly (757)　　　　| 1882—Noble (899)

Mr JOHN HAMAR, Hopton Castle, Aston-on-Clun, Shropshire.

1878—Lord Kilburn (149)
    Marshall (824)
1879—Marshall (824)
1880—Bach's Ram (204)
    Young Kilburn (1311)
1881—Blood Royal (261)
    Minton's No. 9, 1881
    (851)

1881—Minton's No. 25, 1881
    (855)
1882—Blood Royal (261)
    Dawes' Clinker (491)
    Minton's No. 9, 1881 (851)
    Minton's No. 13, 1882
    (857)
    Young Royal (1324)

Mr. JOHN HARDING, Wootton, Bridgnorth, Shropshire.

1870—Marquis (820)
1871—Marquis (820)
1872—Marquis (820)
1873—Marquis (820)
    The Buck (1178)
1874—Nonpareil 3rd (910)
1875—Nonpareil 3rd (910)
1876—Mansell's No. 16, 1876
    (802)
    Sheldon's No. 3, 1876
    (1063)
1877— Mansell's No. 16, 1876
    (802)
    Shamrock (1055)
1878—His Lordship (669)
    Mansell's No. 16, 1876
    (802)
    Pride of Montford (959)

1879—Caligula 2nd (322)
    Mansell's No. 16, 1876
    (802)
    Mansell's No. 15, 1879
    (1225)
    Pride of Montford (959)
1880—Caligula 2nd (322)
    Pride of Montford (959)
1881—Caligula 2nd (322)
    Commissioner (421)
    Primus (962)
    Sir Evelyn (1090)
1882—Cockney (401)
    Commissioner (421)
    Simon de Montford (1089)
    Symmetry (1171)
    The Lawyer (1214)
    The Patriot Lord (1199)

Mr. EDMUND HAWKINS, Dinthill, Ford, Shrewsbury.

1877—Crane and Tanner's No.
    7, 1877 (472)
1878—Crane and Tanner's No.
    7, 1877 (472)
1879—Crane and Tanner's No.
    7, 1877 (472)
    Crane and Tanner's No,
    16, 1879 (475)
1880—Crane and Tanner's No.
    16, 1879 (475)
    Mansell's No. 16, 1876
    (802)

1880—Minton's No. 20, 1880
    (849)
1881—Minton's No. 20, 1880
    (849)
    Minton's No. 9, 1881
    (851)
    Simon de Montford 1089)
1882—Albert (181)
    Ensdon Reserve (526)
    Minton's No. 20, 1880
    (849)

CAPTAIN JOSEPH BOARDMAN HAYDOCK, Wootton Hall,
Henley-in-Arden.

1879—Wootton (1279)
1880—Wootton (1279)
1881—Coxon (458)
        Pulley (975)

1881—Rugby (1033)
1882—Coxon (458)
        Rugby (1033

REV. EDWARD HENSLEY, Parkham Rectory, Bideford,
N. Devon.

1882—Colossus 3rd (412)

CAPTAIN F. HERBERT, Clytha, Usk, Monmouthshire.

1882—Bromfield (298) | 1882—Hubert (675)

MR. W. PYBUS HORNE, Moulton Farm, Richmond, Yorks.

1881—Lord Minton (761) | 1882—Lord Montford (763)

MR. E. J. HULME, Sandon, Stone, Staffordshire.

1882—The Hero (1189)

MR. W. HUMPHREYS, Evenall, Oswestry, Shropshire.

1881—First Flight (563)
1882—First Flight (563)

1882—M.P. (880)

THE EXECUTORS OF THE LATE REV. G. INGE, Thorpe Hall,
Tamworth.

1880—Odstone (919)
1881—Odstone (919)

1882—Odstone (919)
        Pioneer (937)

MR. EDWARD INSTONE, Bourton Grange, Much Wenlock,
Shropshire.

1882—Prince (964)

MR. SKELTON JEFFERSON, Preston Hows, Whitehaven.

1878—Midland (841)
1879—Lothair (779)
        Midland (841)
        Sir Harry (1099)
1880—Calder (320)
        Shipston (1075)
        Sir Harry (1099)

1880—True Form (1241)
1881—Ready Money (993)
        Shipston (1075)
        Sir Matthew (1104)
1882—Battus (220)
        Ready Money (993)

Mr. RICHARD JONES, Jun., Norton, Condover, Shrewsbury

1873—Mansell's Ram (817)
1875—Evans' Ram (537)
    Mansell's Ram (817)
1876—Evans' Ram (537)
    Everall's Ram (538)
    Mansell's Ram (817)
1877—Evans' Ram (537)
    Everall's Ram (538)
1878—Bowen-Jones' Ram (685)

1879—Bowen Jones' Ram (685)
1880—Bromley's Ram (300)
    Duke of Marlboro' (514)
1881—Bromley's Ram (300)
    Duke of Marlboro' (514)
1882—Bromley's Ram (300)
    Everall's No. 16, 1882 (539)
    Fascinator (549)

Mr. R. M. KNOWLES, Colston Bassett, Bingham, Notts.

1878—Salisbury (1038)
1880—Bulwark (308)
1881—Bulwark (308)
    Little Minton (719)
    Town Councillor (1231)

1882—Bulwark (308)
    Colston (413)
    Little Minton (719)
    Magnum (789)
    Multum (883)

Messrs. H. LEE & SON.—*(See Mr. Henry Lee.)*

Mr. HENRY LEE, Ensdon, Montford Bridge, R.S.O., Shropshire.

1875—Crane's No. 3, 1875 (469)
    Mansell's No. 7, 1875 (800)
1876—Genius (600)
    Lord Taunton (121)
1877—Mansell's No. 28, 1877 (805)
    Young Caligula (1293)

1878—Young Caligula (1293)
1879—Son of Columbus (1131)
    Young Caligula (1293)
1880—Royal Reserve (159)
    Son of Columbus (1131)
1881—J. B. Jones' No. 6, 1881 (684)
    Thomas' No. 4, 1881 (1221)

At this period the above Flock in its entirety came into the possession
of
Messrs. H. LEE & SON.

1882—Grand Master (621)
    Thomas No. 4, 1881 (1221)
Young Reserve (1323)

The Right Hon. LORD LEIGH, Stoneleigh Abbey, Kenilworth.

1877—Monmouth (873)
1878—Monmouth (873)
1879—Monmouth (873)
1880—Monmouth (873)
    Young Manager (1314)
1881—Nock's No. 7, 1881 (906)
    Monmouth (873)

1881—Wellington Hero (1270)
    Young Manager (1314)
1882—Nock's No. 7, 1881 (906)
    Williams No. 5, 1882 (1274)
    Young Manager (1314)

Mr. R. LODER, M.P., Whittlebury, Towcester.

1880—Chesham 2nd (362)
Royal Victor (1030)
1881—Chesham 2nd (362)
Dudmaston Hero (165)
Royal Victor (1030)
1882—Bristol Prince (286)
Chesham 2nd (362)

1882—Dudmaston Hero (165)
Earl of Leicester (171)
Lord Chancellor (740)
Royal Victor (1030)
Victor 2nd (1257)
Victor 3rd (1258)
Young Alderman (1288)

Mr. HENRY LOVATT, Low Hill, Bushbury, Wolverhampton.

1875—Beach's No. 1, 1875 (222)
Beach's No. 12, 1875 (223)
1876—Chesham's No. 10, 1876 (366)
Mansell's No. 8, 1876 (801)
Pulley's No. 1, 1876 (979)
1877—Beach's No. 19, 1876 (225)
Challenger (348)
Cossack (445)
Coxon's No. 6, 1876 (460)
1878—Beach's No. 9, 1877 (229)
Chesham's Ram (380)
Cossack (445)
Grandeur (133)
Mansell's No. 20, 1878 (806)
Sheldon's Ram (1071)
1879—Britannicus (287)

1879—Cantab (327)
Commissioner (421)
Grandeur (133)
Mansell's No. 20, 1878 (806)
1880—Byrd's No. 36, 1880 (315)
Commissioner (421)
Grandeur (133)
Mansell's No. 20, 1878 (806)
Mansell's No. 7, 1880 (809)
1881—Commissioner (421)
Williams' No. 4, 1881 (1276)
1882—Mansell's No. 7, 1880 (809)
Mansell's No. 15, 1882 (815)
Royal Gem (1024)

Mr. EDMUND LYTHALL, Radford Hall, Leamington

1852—Aylesford (196)
1853—Old B (920)
1854—Old B (920)
1857—Chester Billy (7)
1863—Alexander (183)
Packington Duke (928)
1864—Packington Duke (928)
1866—Sampson (1042)
1867—Southport (1150)
The Model (1194)
1868—Radford (989)
1869—Corsair (57)

Southport (1150)
Standard Bearer (80)
1870—Warrior (1264)
1873—Earl of Evesham (520)
Son of Islander (1137)
Warrior (1264)
1874—Leviathan (708)
Lord Radford (773)
Tory Peer (1229)
1875—Practical (945)
Premier (948)
1876—Improver (677)

1876—Practical (945)
Premier (948)
Severus 2nd (1051)
The Grecian (1187)
1877—British Prince (291)
Severus 2nd (1051)
1878—British Prince (291)
Burdett Coutts (310)
1879—British Prince (291)

1879—Burdett Coutts (310)
Tribune (1235)
1880—Giant (605)
Triumph (1237)
1881—Richmond (1004)
1882—Alpine (186)
Benedict (244)
Loadstone (722)
Monitor (868)

MR. THOMAS MANSELL, Harrington Hall, Shifnal, Shropshire.

1865—Earl Plymouth (48)
1866—Conservative (435)
1867—Conservative (435)
Mansion 2nd (55)
1868—Conservative (435)
Marquis (820)
1869—Conservative (435)
Marquis (820)
1870—Conservative (435)
Hermit (665)
Marquis (820)
Pattern (933)
Rifleman (1007)
1871—Calcot (317)
Longitude (729)
Pattern (933)
Rifleman (1007)
1872—Calcot (317)
Count (448)
Longitude (729)
Pattern (933)
Severn (1050)
1873—Calcot (317)
Landseer (700)
Pattern (933)
Truestock (1243)
1874—Calcot (317)
The Knight (1190)
Pattern (933)
The General (1186)
1875—Artist (194)

1875—Calcot (317)
The Knight (1190)
Raby Duke (108)
1876—Artist (194)
Calcot (317)
Ruby Duke (108)
Young Calcot (1292)
1877—County Member (452)
Marquis of Bath (822)
Raby Duke (108)
Truestock (1243)
Young Calcot (1292)
1878—Beach's No. 13, 1876 (226)
County Member (452)
North Star (913)
Raby Duke (108)
1879—County Member (452)
Multum in Parvo (885)
North Star (913)
1880—County Member (452)
Lord Clive (742)
Third Marquis of Bute (132)
Multum (882)
1881—Colston (413)
His Lordship 2nd (670)
Lord Clive (741)
The Patriot (1193)
1882—Lord Clive (742)
The Patriot (1198)

The following Rams have been used in the foregoing Flock, but the dates of service are not given.

Courtier (453)
Lord Clifden (44)
Maccaroni (785)
Mansell's No. 6, 1859 (797)

Shropshire Ram (1084)
Short Legged Patentee (1076)
Young Buckskin (1291)

Mr. THOMAS JAMES MANSELL, Dudmaston Lodge, Bridgnorth, Shropshire.

1872—Severn (1050)
Calcot (317)
1873—Calcot (317)
Truestock (1243)
Landseer (700)
1874—Double B (500)
Son of Conservative (1134)
Son of True Type (1148)
True Stock (1243)
1875—Double B (500)
May Duke (837)
Son of True Type (1148)
True Stock (1243)
1876—Beach's No. 17, 1876 (228)
Double B (500)
May Duke (837)
1877—County Member (452)
Double B (500)
Shamrock (1055)
1878—County Member (452)
Double B (500)

1878—His Lordship (669)
Marquis of Bath (822)
Pride of Montford (959)
1879—Character (357)
Milton (843)
Multum-in-Parvo (885)
Pride of Montford (959)
Warwick (1268)
1880—Caligula 2nd (322)
Colston (413)
Milton (843)
Pride of Montford (959)
The Earl (1184)
1881—Commissioner (421)
Sir George (1095)
The Patriot (1198)
Warwick (1268)
1882—Baronet (216)
Commissioner (421)
Disappointment (499)
M.P. (881)
The Patriot (1198)
Warwick (1268)

The Late Mr. T. LOCKLEY MEIRE, Eyton-on-Severn, Shrewsbury.

1878—Shropshire Ram (1083) | 1879—Shropshire Ram (1086)

Mr. EDWARD MEREDITH, Rednal, West Felton, R.S.O., Shropshire.

1869—Montford (869)
1870—Montford (869)
1872—Crane's No. 8, 1872 (467)
Long Back (727)

1873—Crane's No. 8, 1872 (467)
Long Back (727)
Royal Legatee (1026)
1874—The Tory (1208)

1874—Leviathan 2nd (711)
1875—Leviathan 3rd (712)
　　　The Tory (1208)
1876—Leviathan 3rd (712)

1876—The Tory (1208)
1877—Chief (383)
　　　Leviathan 3rd (712)
1878—Chief (383)

At this period the above flock in its entirety came into the possession of

Mr. RICHARD MEREDITH, Rednal, West Felton, R.S.O. Shropshire.

1879—Lord Exeter (745)
1880—British Yeoman (294)
1881—British Yeoman (294)
　　　Commerce (417)

1881—Son of Bristol Reserve (1124)
1882—British Yeoman (294)
　　　Magnum Bonum (790)

Mr. THOS. HORROCKS MILLER, Singleton Park, Poulton-le-Fylde, Lancashire.

1873—Coxon's Critic (459)
1874—Coxon's Critic (459)
1875—Bradburne's No. 26, 1875 (267)
1876—Bradburne's No. 26, 1875 (267)
1877—Duke of Hattons (512)
1878—Duke of Hattons (512)

1879—Baron Uffington (219)
　　　Breastplate (270)
1880—Breastplate (270)
　　　Mountaineer (877)
1881—Bishton (258)
1882—Lord of the Harem (767)
　　　Monarch (867)

Mr. J. W. MINTON, Forton, Shrewsbury.

1880—Forton 1st (573)
1881—Forton 1st (573)

1882—Forton 1st (573)
　　　Forton 2nd (574)

Mr. THOMAS STEPHEN MINTON, Montford, R.S.O., Shropshire.

1873—Son of Conservative (1134)
　　　Son of Little Lord (1139)
1874—Bedford Hero (111)
　　　Son of Conservative (1134)
　　　Son of Little Lord (1139)
1875—Bedford Hero (111)
1876—Son of Bedford Hero (1123)
　　　Son of Calcot (1125)
1877—Dudmaston (504)
　　　Pride of Montford (959)

1877—Son of Bedford Hero (1123)
1878—His Lordship (669)
　　　Marquis of Bath (822)
1879—Marquis of Bath (822)
1880—Ercall P. alias Son of Pride of Montford (528)
　　　Milton (843)
　　　Montford Hero (160)
　　　Shipston (1074)
1881—His Lordship 2nd (670)
　　　Thickset (1214)
1882—Baron Plassy (218)
　　　Minton's Pride (858)

Mr. P. ALBERT MUNTZ, Dunsmore, Rugby.

1882—Latimer Wonder (703) | 1882—Pride of Dudmaston (956)

Mr. J. LENOX NAPER, Loughcrew, Oldcastle, Ireland.

1865—Negro (888)
1866—Negro (888)
1868—Julius (690)
    Speculum (1154)
1869—Duke of Edinboro' (511)
    Julius (690)
    Speculum (1154)
1870—Julius (690)
1871—Lord Haughton (751)
1872—Marshall Freeford (825)
    Viceroy (1256)
1873—Paddy Green (930)
    The Proctor (1203)
1874—Conway (440)
    Reformer (999)
1875—Conway (440)
    Longbow (728)
    Reformer (999)

1876—Cossack (445)
    Longbow (728)
    The Baron (1177)
1877—Cossack (445)
    Longbow (728)
    Sir Gray (1096)
    The Baron (1177)
1878—Cossack (445)
    Monarch (865)
    Quality (982)
1879—Monarch (865)
    Quality (982)
1880—Protector (972)
    Quality (982)
1881—Prince Regent (970)
    Protector (972)
1882—Prime Regent (970)

Mr. RYDER J. NASH, Park House, Glanmire, Co. Cork, Ireland.

1872—Royal Blood (1020)
    Sheldon's No. 2, 1872 (1062)
1873—Royal Blood (1020)
    Sheldon's No. 2, 1872 (1062)
1874—Pulley's No. 9, 1874 (978)
    Sheldon's No. 2, 1872 (1062)
1875—Sheldon's No. 2, 1872 (1062)
    Pulley's No. 9, 1874 (978)
    Royal Reserve (1027)
1876—Major (792)

1876—Pulley's No. 9, 1874 (978)
    Truisbe (1244)
1877—Major (792)
    Truisbe (1244)
1878—Nobleman (901)
    Truisbe (1244)
1879—Noble (898)
    Nobleman (901)
1880—Brailes (268)
    Hard Times (643)
1881—Hard Times (643)
1882—Duneske (516)
    Hereford (664)

H

THE LATE MR. V. E. NIGHTINGALE, Burway, Ludlow, Shropshire.

1858—Bach's Ram (203)
1860—Henry Smith (661)
    Son of Bach's Ram (1122)
1861—Hand's Ram (643)
1862—Sheldon's Ram (1067)
    Son of Henry Smith (1135)
1863—Sheldon's Ram (1067)
1864—Sheldon's Ram (1067)
    Hand's Ram (644)
    Worcester (1280)
1867—Son of Worcester (1149)
1870—Mansell's No. 2 (808)
1873—Byrd's No. 4 (313)

1875—Chesham's No. 9, 1875 (365)
1876—Favourite (554)
1877—General (595)
    Grand Duke (620)
1878—Carouser (341)
1879—Downton Pippin (503)
1881—Burway Crane (311)
    Burway Pippin (312)
    Royalist (173)
1882—Burway Crane (311)
    Crane and Tanner's No. 4, 1882 (478)

MR. THOMAS NOCK, Sutton Maddock, Shifnal, Shropshire.

1870—Blood Royal (83)
1871—Blood Royal (83)
1872—Blood Royal (83)
1873—Blood Royal (83)
1874—Blood Royal (83)
1875—Blood Royal (83)
    Defiance (497)
    True Light (1242)

1876—Sandboy (1044)
1877—Sandboy (1044)
1878—Sandboy (1044)
    Talisman (136)
1879—Sandboy (1044)
    Talisman (136)
1880—Sandboy (1044)
    Talisman (136)

MR. EDWARD NOCK, Brockton House, Shifnal, Shropshire.

1876—Aston (125)
1877—Aston (125)
1878—Aston (125)
    Cynic (484)
1879—Aston (125)
    Cynic (484)
1880—Aston (125)
    Bristol Reserve (141)
    Cynic (484)
1881—Aston (125)

1881—Bristol Reserve (141)
    Lord Clive (742)
    Southport Prize (1151)
    Star (1160)
1882—Aston (125)
    Bristol Reserve (141)
    Lord Clive (742)
    Ultimatum (1251)
    Young Cockade (1300)

CAPT. ROBERT E. OLIVER, Sholebrook Lodge, Towcester.

1870—Shropshire Ram (81)
1871—Beach's Ram (235)
1872—Pulley's Ram (976)

1873—Pulley's Ram (976)
    Pulley's Ram (977)
1874—Sheldon's Ram (1068)

1875—Townshend's Ram (1232)
1876—Sheldon's Ram (1069)
1877— Smith's Ram (1119)
1878—Chesham's Ram (379)

1879—Loder's Ram (725)
1881—Williams' Ram (1275)
1882—Loder (726)

Mr. J. McLEOD PETLEY, The Green House, Bridgnorth.

1881—Duke Goliah (508)
    Lord Oxon (167)
    Primus (962)
    Sir Evelyn (1090)
1882—Cockney (401)

1882—Duke Goliah (508)
    Lord Oxon (167)
    Sir Evelyn (1090)
    Symmetry (1171)
    The Lawyer (1214)

Mr. JOHN PICKERING, Alston, Shrewsbury.

1867—Duke of Leinster (513)
1868—Duke of Leinster (513)
1869—Duke of Leinster (513)
    Young Marquis (1315)
1870—Duke of Leinster (513)
    Young Marquis (1315
1871—Young Marquis (1315)
1872—Young Marquis (1315)
1873—Patent (932)
1874—Patent (932)

1875—Patent (932)
1876—Patent (932)
1879—Thomas No. 1, 1879 (1217)
1880—Pride of Oxon (960)
    Thomas No. 1, 1879 (1217)
1881—Cœur-de-Lion (404)
    Pride of Baschurch (953)
1882—First Choice (562)
    Pride of Baschurch (953)

Mr. SAMUEL C. PILGRIM, The Outwoods, Burbage, Hinckley

1877—Julius (691)
    Matchless (834)
1878 —Cynosure (485)
    Julius (691)
    Matchless (834)
1879—Matchless (834)
    Normanton (912)

1879—Victor (150)
1880—Alderman (182)
    Shelford (1072)
    Victor (150)
1881—Punch (981)
    Victor (150)
1882—Punch (981)

The Right Hon. LORD POLWARTH, Humbie House, Upper Keith, C. Lothain.

1872—Son of Standard Bearer
    (1147)
1873—Son of Standard Bearer
    (1147)
1874—Son of Standard Bearer
    (1147)

1875— Grand Vizier (630)
    Son of Oxford Hero (1142)
1876—Grand Vizier (630)
    Son of Oxford Hero (1142)
    The Young Sultan (1212)
1877—Grand Vizier (630)

1877—Son of Oxford Hero(1142)
1878—Ironmaster (682)
  Son of Oxford Hero(1142)
  Young Masterman (1316)
1879—Ironmaster (682)
  Young Masterman (1316)
1880—Young Masterman (1316)

1881—Young Masterman (1316)
1882—Mansell (796)
  Young Lord Falmouth (1312)
  Young Montford Hero (1318)

This Ram has been used in the above Flock, but no date of service given.

Lord Stanley (776)

Mr. FRANK POVEY, Lee, Ellesmere, Shropshire.

1878—Clinker (398)
1879—Clinker (398)
1880—Rednal Chief (997)

1881—Rednal Chief (997)
1882—Merry Monarch (84)

Mrs. A. PRESCOTT, Birches Farm, Tenbury, Worcestershire.

1881—Baron (211)
1882—Circuit (390)

1882—Non-Such (911)

Mr. GEORGE CORSER PRICE, Acton Hill, Stafford.

1881—Robin Hood 2nd (1015)
1882—Robin Hood 2nd (1015)

1882—Little John (717)

Mr. J. PULLEY, M.P., Lower Eaton, near Hereford.

1870—Fat Back (551)
1871—Buckskin (307)
  Proud Salopian (87)
1873—Sultan (1167)
1874—Shrewsbury (1081)
1875—Grand Duke (620)
1876—Young Sultan (117)
1877—Young Sultan (117)

1877—Young Colossus (1301)
1878—Cannock Chief (326)
  Young Colossus (1301)
  Young Sultan (117)
1880—Sir Roger (1112)
1881—Reality (994)
1882—Sultan 3rd (1168)

Mr. CHARLES RANDELL, Chadbury, near Evesham.

1872—Copperplate (441)
  Duke of Cambridge (509)
  Field Marshall (559)
  Our Chief (924)

1872—Typical (1249)
1873—Copperplate (441)
  Duke of Cambridge (509)
  Field Marshall (559)

1873—Our Chief (924)
     Typical (1249)
1874—Chieftain (386)
     Copperplate (441)
     Corporal (444)
     Field Marshall (559)
     Our Chief (924)
     Typical (1249)
1875—Chieftain (386)
     Corporal (444)
     Our Chief (924)
     Typical (1249)
1876—British Yeoman (295)
     Corporal (444)
     Defiance (497)
     Duke of Cornwall (510)
     Our Chief (924)
     Prince Charlie (965)
1877—Colossal (410)
     Defiance (497)
     Duke of Cornwall (510
     Jack Tar (683)
     Julius Cæsar (692)
     L. S. D. (781)
     Master Beach (826)
     Nelson (889)
     Our Chief (924)
     Prince Charlie (965)
     R. N. (1011)
1878—A1 (177)
     Colossal (410)

1878—Defiance (497)
     Jack Tar (683)
     Julius Cæsar (692)
     L. S. D. (781)
     Master Beach (826)
     Nelson (889)
     R. N. (1011)
1879—British Chieftain (288)
     Colossal (410)
     Julius Cæsar (692)
     Quarryman (986)
     Sutton Maddock (1169)
1880—Big Corporal (254)
     Capital (330)
     Colossal (410)
     Gigantic (606)
     Quarryman (986)
     Sir Henry (1100)
1881—Big Corporal (254)
     Capital (330)
     Collingwood (407)
     Colossal (410)
     D.D. (493)
     Gigantic (606)
1882—Capital (330)
     Collingwood (407)
     Commodore (420)
     D.C. (492)
     D.D. (493)
     Lord Alcester (732)
     Royalist (173)

COLONEL ALEXANDER RIDGWAY, Sheplegh Court,
Blackawton, South Devon.

1882—Bristol Pet (284)
     Lord Milcombe (759)

1882—The Sultan (1207)

MR. JOHN RILEY, Putley Court, Ledbury, Herefordshire.

1880—Restorer (1003)
1881—Restorer (1003)

1882—Restorer (1003)
     Sextus (1053)

MISS GERTRUDE ROSE, Mullaghmore, Monaghan, Ireland.

1882—Broad Back (297)
     King William (699)

1882—Latimer (702)

Mr. THOMAS RYLAND, Pheasey Farm, Queeslet, Perry Barr, Birmingham.

1872—Radical (990)
1873—Radical (990)
1874—Rufus (1032)
1875—Righton (1010)
    Rowland (1031)
1876—Rector (995)
    Robin (1013)
    Rowland (1031)
1877—Czar (489)
    Hopeful (671)
    Rogue (1019)
1878—Czar (489)

1878—Hopeful (671)
    Lord Exeter (745)
    Ridley (1006)
1879—Brigadier (276)
    Hopeful (671)
    Ridley (1006)
    Sir William (1115)
1880—Concord (427)
    Congreve (431)
    Ruler (1034)
    Young Hopeful (1310)

At this period the above flock in its entirety came into the possession of

Mr. A. S. BERRY, Pheasey Farm, Queeslet, Perry Barr, Birmingham.

1881—Antique (189)
    Arley (193)
    Concord (427)
    Congreve (431)
    Young Hopeful (1310)
1882—Advancer (180)

1882—Anchor (188)
    Concord (427)
    Lady Foppington (748)
    The Colonel (1180)
    Young Hopeful (1310)

Mr. HENRY J. SHELDON, Brailes House, Warwickshire.

1854—Bach's Ram (202)
    Juckes' Ram (689)
1855—Bach's Ram (202)
    Juckes' Ram (689)
1856—Attraction,(195)
    Foster's Ram (578)
1857—Attraction (195)
    Foster's Ram (578)
1858—Brother to Earl of Warwick (302)
    Shropshire Ram (5)
1859—Brother to Earl of Warwick (302)
    Fat Back (550)
1860—Brother to Earl of Warwick (302)
    Lord Salisbury (1192) *See Errata.*

1861—Foster Leeds H. C. Ram (577)
    Lord Astley (735)
    Young Magnum Bonum (1313)
1862—Black Prince 2nd (26)
    Lord Astley (735)
1863—Lord Astley (735)
    Quality (32)
    Coventry (456)
1864—Mansell's Newcastle Ram (816)
1865—Coventry (456)
    Horton's Big Ram (672)
1867—Young Perfection (1321)
1868—Shropshire Ram (62)
    Young Perfection (1321)

1869—Evans' No. 7, 1869 (531)
Young Perfection (1321)
1870—Horley's No. 29, 1870 (673)
Tory Peer, (1229)
1871 Tory Peer (1229)
1872—Calcot (317)
Tory Peer (1229)
True Type (85)
1873—Goliah (611)
Tory Peer (1229)
1874—Goliah (611)
Lord Chesham's Nock (378)
1875—Goliah (611)
Lord Chesham's Nock (378)
1876—Beach's No. 9,1876 (224)
Philistine 936)
Taunton Reserve (118)
1877—Philistine (936)
Taunton Reserve (118)
1878—Generalissimo (599)
Model (862)
Sheldon's Kilburn Ram (1061)
Sheldon's No. 1, 1879 (148)
1879—Beaconsfield (237)
British Yeoman (294)
Lord Blaydon (738)

1879—Model (862)
1880—Beach's R.A.S. Ram (230)
Graham's No. 6, 1880 (614)
Graham's No. 7, 1880 (615)
Lord Blaydon (738)
Model (862)
1881—Chesham's No. 26, 1881 (371)
Chesham's No. 44, 1881 (372)
Georgius (139)
Graham's No. 7, 1881 (616)
Graham's No. 10, 1881 (617)
Model (862)
1882—Chesham's No. 26, 1881 (371)
Chesham's No. 44, 1881 (372)
George the Fourth (601)
Graham's No. 7, 1881 (616)
Graham's No. 10, 1881 (617)
Sheldon's A. (1056)
Sheldon's B. (1057)
Sheldon's C. (1058)
Sheldon's D. (1859)

THE LATE RIGHT HON. THE EARL OF SHREWSBURY,
The Birches, near Rugeley.

1872—Scottish Chief (1045)
1873—Hercules (662)
1874—Hercules (662)
1875—Conqueror (433)

1875—Sir William (1114)
The Moor (1195)
1876—Sir William (1114)

At this period the above Flock in its entirety came into the possession
of
THE RIGHT HON. THE EARL OF SHREWSBURY.

1877—Forton Hero (575)
M.A. (784)
Minton's No. 12, 1877 (846)

1877—Monarch (864)
1878—Boreas (263)
Forton Hero (575)
1878—Minton's No.12, 1877 (846)

1879—Minton's No. 12, 1877 (846)
Mitre (861)
The Star (1206)
1880—Brereton Boy (271)
Grand Seigneur (626)
Mitre (861)

1881—Blue Jacket (262)
Grand Seigneur (626)
Mystery (887)
1882—Grand Seigneur (626)
The Councillor (1181)
The Noble (1196)

THE RIGHT HON. THE EARL OF STRATHMORE, Glamis Castle, Forfarshire.

1869—Scottish Hero (1046)
1870—Standard Bearer (80)
1871—Standard Bearer (80)
1872—Standard Bearer (80)
1873—Bedford Hero (111)
1875—Young Bedford Hero (1290)
1876—Young Bedford Hero (1290)
1877—Mansell's No. 4, 1877 (1224)

1878—Mansell's No. 4, 1877 (1224)
1879—Forester (570)
Mansell's No. 4, 1877 (1224)
1880—Forester (570)
Lord Brailes (739)
1881—Lord Brailes (739)
Sir George (1094)
1882—Lord Brailes (739)
Prototype (971)
Sir George (1094)

MR. RICHARD THOMAS, The Buildings, Baschurch, Shropshire.

1860—Mansell's No. 5, 1860 (1226)
1861—Mansell's No. 5, 1860 (1226)
1862—Mansell's No. 5, 1860 (1226)
1863—Mansell's No. 5, 1860 (1226)
The Ruler (1204)
1864—De Broke (494)
The Ruler (1204)
Young Clifden (1299)
1865—De Broke (494)
The Ruler (1204)
Young Clifden (1299)
1866—De Broke (494)
Young Clifden (1299)
1867—De Broke (494)
1868—Mansell's No. 6, 1868 (798)

1869—Calcot Chieftain (319)
Mansell's No. 6, 1868 (798)
1870—Castle Warden (342)
Mansell's No. 6, 1868 (798)
1871—Co-Monument (423)
Favourite (553)
Mansell's No. 6, 1868 (798)
1872—Benthall Chieftain (246)
Favourite (553)
Foreman (568)
1873—Benthall Chieftain (246)
Favourite (553)
Foreman (568)
Son of Conservative (1133)
1874—Benthall Chieftain (246)
Foreman (562)
1875—Prince (963)

1876—Prince (963)
1877—Grandeur (133)
    Prince (963)
1878—Claudius 3rd (396)
    Grand Chief (618)
    Prince (963)
1879—A1 (176)
    Grand Chief (618)
1880—A1 (176)
    Cœur-de-Lion (404)
    Grand Chief (611)

1880—Grand Master (621)
    Standard (1159)
1881—Cœur-de-Lion (404)
    Grand Master (621)
    Prince Royal (971)
    The Patriot (1198)
1882—A. A. (175)
    Grand Prince (625)
    Protector (973)
    Safeguard (1036)

MR. CHARLES TIMMIS, Gainsboro' Hill, Stonnall, Walsall.

1873—Earl of Evesham (520)
1874—Earl of Bedford (518)
1875—Lord Apley (734)
1877—Monarch (863)
1878—Bendigo (243)
    Monarch (863)
    Statesman (1161)

1879—Bendigo (243)
    Monarch (863)
    Richmond (1004)
    Statesman (1161)
1880—Cyrus (487)
1881—Falstaff (541)
1882—Connaught (432)

MR. ROBERT TIMMIS, Dryton, Wroxeter, Shropshire.

1868—Lord Milford (760)
1870—Straight Back (1164)
1871—Mountain Hero (878)
1873—Grand Master (622)
1875—Trooper (1245)
1876—Byron (316)
1877—Newark (891)
1878—Newport John (892)

1881—Chesham's No. 8, 1881
    (370)
    Harding's No. 2, 1881
    (646)
    Sheldon's No. 2, 1881
    (1066)
1882—Chesham's No. 1, 1882
    (373)

CAPTAIN HENRY TOWNSHEND, Caldicote Hall, Nuneaton.

1875—Baron (210)
1876—Christy (388)
1877—Freeman (585)
1878—Brother to Lord Liver-
    pool (303)
1879—Congress (430)

1879—Tartar (1174)
1880—Cumberland Hero (155)
1881—Sir George (1093)
    Taunton Duke (1175)
1882—Notable(914)
    Sir George (1093)

Mr. ARTHUR P. TURNER, Strangworth, Pembridge, Herefordshire.

1876—Chesham's No. 20, 1876
(369)
1877—Chesham's No. 20, 1876
(369)
1878—Chief (382)
Farmer's No. 15, 1878
(544)
1879—Chief (382)
Farmer's No. 15, 1878
(544)

1880—Chief (382)
Farmer's No. 10, 1880
(545)
1881—Farmer's No. 10, 1880
(545)
1832—Farmer's No. 10, 1880
(545)
Forester (571)

Sir HENRY HUSSEY VIVIAN, BART., M.P., Park-le-Breos, Swansea, South Wales.

1880—Farmer's No. 16, 1880
(546)
Nightingale No. 1, 1880
(895)

1880—Nightingale No. 2, 1880
(896)
1881—British Flag (289)
1882—Tornado (1228)

The Late Mr. CHARLES WADLOW, Haughton, Bridgnorth, Shropshire.

1879—Warwick (1268)
1880—Bridgnorth (274)
Warwick (1268)

1881—Bridgnorth (274)
1882—Bridgnorth (274)

Mr. EDMUND C. WADLOW, Stanton Hill, Shifnal, Shropshire

1879—Nugget (916)
1880—Nugget (916)
1881—Fraternity (581)
Nugget (916)

1882—Fraternity (581)
Nugget (916)
Precocity (946)

Mr. T. WALKER, Berkswell Hall, Meriden, Coventry.

1881—Grindle 2nd (641)
Loder's No. 1, 1881 (723)

1882—Grindle 2nd (641)
Harlescott 2nd (653)
Loder's No. 1, 1881 (723)

Mr. WILLIAM WARD, Shrawardine Castle, Montford Bridge, R.S.O., Shropshire.

1871—Chesham's No. 10, 1871 (363)
1872—Chesham's No. 10, 1871 (363)
1873—Foreman (568)
1874—Calcot Chieftain (319)
　　　Foreman (568)
1875—Shropshire Ram (1085)
1876—Crane's No. 12, 1876 (471)

1877—Crane's No. 12, 1876 (471)
1878—Minton's No. 1, 1878 (847)
　　　Warden (1263)
1879—Son of Prince (1145)
　　　Williams' No 1, 1879 (1273)
1881—Active (178)
　　　Prince of Wales (969)
1882—Grand Master (621)

Mr. JAMES WATSON, Berwick House, Shrewsbury.

1881—Flag Staff (566)
　　　Warrior (1266)
1882—Baron Alkmond (214)

1882—Berwick (250)
　　　Berwick Hero (251)

Mr. GEORGE W. WHEELER, Posenhall, Broseley, Shropshire.

1873—Posenhall (942)
1874—Chesham's No. 19 (364)
　　　Posenhall (942)
1875—Smith's No. 12, 1875 (1118)
1876—Smith's No. 12, 1875 (1118)
1877—Brutus (306)
　　　Smith's No. 12, 1875 (1118)

1878—Brutus (306)
1879—Brutus (306)
　　　His Lordship (669)
　　　Leo (707)
1880—Dudmaston (505)
　　　Posenhall Lordship (943)
1881—Dudmaston (505)
　　　Posenhall Lordship (943)
1882—Bourton (265)
　　　Posenhall Lordship (943)

Mr. MATTHEW WILLIAMS, Senr., Dryton, Wroxeter.

1868—Exile (540)
1872—Cambrian (324)
1873—Co-Monument (422)
　　　Lord Napier (764)
1874—Co-Monument (422)
　　　Nobleman 3rd (902)

1875—Co-Monument (422)
1876—Artist (194)
　　　Duke of Wellington (515)
　　　True Light (1242)

At this period a portion of the above Flock came into the possession of

Mr. MATTHEW WILLIAMS, Bishton Hall, Shifnal, Shropshire.

1877—Bristol Reserve (144)
1878—Sheldon's No. 4, 1876 (1064)

1879—Clinker (397)
　　　Hercules (663)
1880—Harlescott (651)

1880—Hercules (663)
    Pride of Bishton (954)
1881—Beaconsfield (238)
    Harlescott (651)
    Hercules (663)

1881—Pride of Bishton (954)
1882—Beaconsfield (823)
    Harlescott (651)
    Prince Regent (169)

The Late Right Hon. LORD WILLOUGHBY DE BROKE,
Compton Verney, Warwick.

1860—Crafty (464)
    Duke of Kent (13)

1861—Horton's No. 16, 1861
    (674)

At this period the above Flock in its entirety came into the possession

of

The Right Hon. LORD WILLOUGHBY DE BROKE.

1862—Duke of Kent (13)
    Lord of the Isles (769)
1863—Sir Harry (1098)
1864—Capt. Semmes (33)
1865—Quality 3rd (985)
    Young Patentee (1320)
1866—The Young Duke (1211)
1867—Evans' No. 12, 1867 (530)
    The Earl (1183)
1868—Evans' No. 12, 1867 (530)
    Son of Quality (1146)
    Young Duke (1304)
1869—Leviathan (72)
    Young Duke (1304)
    Grandson of Nobleman (628)
1870—Commodore (419)
    Marquis (821)
    Young Duke (1304)
1871—Nonpareil 3rd (909)
1872—Commander (415)
    Leviathan (72)
    Nonpareil 3rd (909)
    Smith's No. 7, (1117)
    Son of Leviathan (1138)
    Young Duke (1304)

1874—Grandson of Oxford Hero (627)
    Son of Commander-in-Chief (1132)
1875—General (598)
    Leviathan (710)
    Lord Hull (100)
    Son of Pattern (1143)
    Son of Commander-in-Chief (1132)
1876—General (598)
    Lord Hull (100)
    Son of Claudius (1129)
1877—Second to None (1048)
    The Laird (1193)
1878—Minton's Prize (859)
    The Laird (1193)
    Thomas' No. 3, 1878 (1216)
1879—Bromley's No. 13, 1879 (299)
    Fenn's No. 3, 1879 (555)
    Minton's Prize (859)
    Thomas' No. 30, 1879
1880—Everall's No. 3, (529)
    Nock's No. 5, (905)
1882—Constantine (439)
    Nock's No. 10 (907)

Mr. H. WILSON, Pendeford, near Wolverhampton.

1881—Freeford Duke (584)
    Lord Hatton (750)
    Stafford Hero (1156)
    Young Salopian (1326)
1882—Bristol Beau (281)

1882—Freeford Duke (584)
    Lord Hatton (750)
    Stafford Hero (1156)
    Young Salopian (1326)

Mr. WALTER WILSON, Wildmoor, Claverley, Bridgnorth, Shropshire.

1879—Snowflight (1121)
1880—Snowflight (1121)
1881—Bumper (309)
    Lord Alfred (733)

1882—Glendower (609)
    Lord Alfred (733)
    Tyndale (1247)

Mr. MYLES WOODBURNE, Kenwick, Shrewsbury.

1876— Chesham's No. 15, 1876 (367)
    Matthew's No. 1, 1876 (835)
1877—Chesham's No. 15, 1876 (367)
    Matthew's No. 1, 1876 (835)
1878—Chesham's No. 15, 1876 (367)
    Matthew's No. 1, 1876 (835)
1879—Harlescott (652)
    Shrawardine (1078)

1880—Harlescott (652)
    Kenwick (694)
    Shrawardine (1078)
1881—Alpha (185)
    Harlescott (652)
    Kenwick (694)
    Shrawardine (1078)
    Thomas No. 19, 1881 (1223)
1882—Alpha (185)
    Black Knight (260)
    Harlescott (652)
    The Poet (1201)
    Victor (1260)

The Right Hon. The EARL OF ZETLAND, Aske, Richmond, Yorkshire.

1878—Stafford (1155)
1879—Post Captain (944)
    Stafford (1155)
1880—Post Captain (944)
    Stafford (1155)
1881—Post Captain (944)
    Shrawardine (1080)

1881—Stafford (1151)
1882—Montford (871)
    Post Captain (944)
    Royal Cardinal (1021)
    Royal Standard (1028)
    Stafford (1155)

APPLICATION for the "FLOCK BOOK," or any particulars respecting the work to be addressed to the Secretaries—

Messrs. LYTHALL & MANSELL,

COLLEGE HILL, SHREWSBURY;

BINGLEY HALL, BIRMINGHAM; &

LICHFIELD.

LIST OF MEMBERS.

LIFE MEMBERS MARKED THUS *

*Adderley, Joseph Robinson, Teddesley Coppice, Penkridge, Stafford.
*Allen, George, Yew Tree Farm, Penkridge, Stafford.
 Allsopp, Sir Henry, Bart., Hindlip Hall, Worcester.

*Bach, Francis, Onibury, Salop.
*Bach, R., Park Farm, Onibury, Salop.
 Baker, William, Moor Barns, Atherstone, Warwick.
*Barber R., Harlescott, near Shrewsbury.
 Barrs, J. A., Nailstone, Hinckley, Leicestershire.
 Barrs, Mrs., Odstone Hall, Atherstone.
*Bassett, C. H., Pilton House, Barnstaple.
*Bateman, Lord, Shobdon Court, Shobdon, Herefordshire.
*Beach, Joseph, The Hattons, Brewood, Staffordshire.
*Bell, G. J., The Nook, Irthington, Carlisle.
 Bennett, Hugh, Builth Wells, Breconshire.
*Berry, A. S., Pheasey Farm, Queeslet, Birmingham.
*Bethune, Alexander, Blebo House, Cupar-Fife.
*Blantern, G. G., Balderton Hall, Myddle, R.S.O., Salop.
 Bourne, J., Arbour Farm, Market Drayton.
*Bowen-Jones, J., Ensdon House, Montford Bridge, R.S.O.
 Bradburne, H. and A., Pipe Place, Lichfield.
*Brown, Joseph, Bednall, Stafford.
*Brown, Richard, Ruyton-XI-Towns, Salop.
*Bromley, R., Felton Butler, Montford Bridge, R.S.O., Salop.
 Burne, S. T. H., Loynton Hall, Newport, Salop.
 Buttar, D., Corston, Couper Angus, Forfar, N.B.
*Byrd, C., Littywood, Stafford.

*Cartwright, T., The Buildings, West Felton, R.S.O.
*Cawdor, Earl of, Stackpole Court, Pembroke
*Chesham, Lord, Latimer, Chesham, Bucks.
*Clare, W. H. L., 33, Friar Lane, Leicester.
*Clark, C. F., Perton Grove, Wolverhampton.
*Coleman, E. M., Beechenhurst, Selly Hill, Birmingham.
 Colchester-Wemyss, M. W., The Wilderness, Mitcheldean, Gloucester-
    shire, R.S.O.
 Cooke, Grimwood, Horse Heath Park, Linton, Cams.

Cotterell, W.. Derry Ormond, Cardiganshire.
*Coxon, C., Elford Park, Tamworth.
Coxon, J., Freeford, Lichfield.
Coxon, J. L., Tempe, Swepston, Ashby-de-la-Zouch.
Cradock, Major R. W., Derrycalaghan, Roscrea, Co. Tipperary.
*Crawshay, H., Stormer Hall, Leintwardine.
*Crawshay, W. T., Cyfartha Castle, Merthr Tydfil, S. Wales.

*Darling, John, Beau Desert, Rugeley.
*Davies, E., Gatacre Park Farm, Bridgnorth.
*Dicken, T., Ellerdine, Wellington, Salop.
Donne, H., Leek Wootton, Warwick.
Douthwaite, W. C., Rousham, Banbury, Oxon.
Dyott, Col., Freeford Hall, Lichfield.

Edwards, George, Little Brampton, via Kington, Herefordshire.
*Edwards, George W., Woolston, Oswestry.
*Edwards, J. W., The Court, West Felton, R.S.O.
Evans, John, Uffington, near Shrewsbury.
*Everall, Peter, Uckington, Wroxeter, Salop

*Falmouth, Viscount, Tregothnan Park, Cornwall.
*Farmer, J. E., Felton, Ludlow.
Fenn, Thomas, Stonebrook House, Downton, Salop.
Fowler, William, Acton Reynald, near Shrewsbury.

*Gibson, A. S., Bulwell Hall, Notts.
Gibson, F., Woolmet, Dalkeith.
German, Exors. of the late W., Measham Lodge, Atherstone.
Green, J. B. & G. H., Marlow Lodge, Leintwardine.

*Hall, Joseph, Birchdale, Frodsham, Cheshire.
*Hamar, John, Hopton Castle, Aston-on-Clun, Salop.
Hancock, T. F., Whittlebury, Towcester.
*Harding, John, Wootton, Bridgnorth, Salop.
*Hawkins, E., Cruckfields, Ford, Salop.
*Haydock, Capt. J. B., Wootton Hall, Henley-in-Arden.
*Headfort, The Marquis of, Headfort House, Co. Meath, Ireland.
Hensley, Rev. E., Parkham Rectory, Bideford, Devon.
*Horne, W. Pybus, Moulton Farm, Richmond, Yorks.

*Inge, W. F., Thorpe Hall, Tamworth.
*Instone, Henry, Cound, Salop.
*Instone, Thomas, Callaughton, Much Wenlock.

Jefferson, Skelton, Preston Hows, Whitehaven.
*Jones, R., Norton, Condover, Salop.

*Knowles, R. M., Colston Bassett, Bingham, Notts.

*Lee, Thomas, Ensdon, Montford Bridge, R.S.O., Salop.
*Lewis, George, Ercall Park, Wellington, Salop.
*Loder, R., M.P., Whittlebury, Towcester.
*Lovatt, H., Low Hill, Wolverhampton.
*Low, Gavin, 50, Prussia Street, Dublin.
 Lythall, E. Radford Hall, Leamington.

*Mann, W. R., Bolarum Leamington.
*Mansell, A. E., Wyken, Worfield, Bridgnorth, Salop.
 Mansell, Thomas, Harrington Hall, Shifnal.
*Mansell, T. J., Dudmaston Lodge, Bridgnorth, Salop.
 Mansfield, Earl of, Scone Palace, Perth.
 Meire, G. H., Eyton-on-Severn, Shrewsbury.
 Meredith, R., Rednal, West Felton, R.S.O., Salop.
 Miller, R. S., Badger, Shifnal, Salop.
*Miller, T. Horrocks, Singleton Park, Poulton-le-Fylde, Lancashire.
*Minton, T. S., Montford, Montford Bridge, R.S.O., Salop.
*Muntz, P. A., Clifton-on-Dunsmore, Rugby.

*Naper, J. L., Loughcrew, Oldcastle, Ireland.
 Nash, R. J., Park House, Glanmire, Co. Cork.
 Nightingale, The late V. E.. Burway, Ludlow.
 Nock, E., Brockton House, Shifnal.
 Norman, W., Hall Bank, Aspatria.

 Oliver, Capt. R. E., Sholebrook Lodge, Towcester.

 Petley, J. McLeod, The Greenhouse, Bridgnorth, Salop.
 Pickering, John, Alston, Lea Cross, Pontesbury, Salop.
 Pilgrim, S. C., The Outwoods, Burbage, Hinckley, Leicestershire.
*Polwarth, Lord, Merton House, Berwickshire.
*Povey, Frank, Lee Farm, Ellesmere, Salop.
 Prescott, Mrs. A., Birches Farm, Tenbury, Worcestershire.
*Price, G. Corser, Acton Hill, Stafford.
*Pugh, W. R., Ewdness, Bridgnorth, Salop.
*Pulley, Joseph M.P., Lower Eaton, Hereford.

 Randell, C. Chadbury, Evesham.
*Ridgway, Col., Sheplegh Court, Blackawton, R.S.O., S. Devon.
 Robinson, Stephen., Lynhales, Kington, Herefordshire.
 Rose, Miss, Mullaghmore, Monaghan, Ireland.

*Scratton, D. R., Ogwell, Newton Abbott.
*Sheldon, H. J., Brailes House, Shipston-on-Stour.
 Sheraton, W., Broom House, Ellesmere, Salop.
*Sherwood, John, Westwood, Brimstage, near Birkenhead.

*Shrewsbury, Earl of, Ingestre Hall, Stafford.
 Smith, F. D., Lea, Halesowen, near Birmingham.
*Strathmore, Earl of, Glamis Castle, Forfar, N.B.

*Tanner, Alfred, Shrawardine, Montford Bridge, R.S.O., Salop.
*Thomas, R., The Buildings, Baschurch, Salop.
 Timmis, C., Gainsboro' Hill, Stonnall, Walsall.
*Timmis, R., Dryton, Wroxeter.
 Towneley-Parker, T. Towneley, Lytham, Lancashire.
*Townshend, H., Caldecote Hall, Nuneaton.
 Turner, A. P., Strangworth, Pembridge, Herefordshire.

 Vivian, Sir Henry H. Bart., M.P., Park-le-Breos, Swansea, S. Wales.

*Wadlow, E. C., Stanton Hill, Shifnal.
 Ward, William, Shrawardine Castle, Montford Bridge, R.S.O., Salop.
*Warwick, Earl of, Castle Park, Warwick.
*Watson, James, Berwick Hall, Shrewsbury.
*Webb, Messrs. E. & Sons, Wordsley, Stourbridge.
*Wheeler, George W., Posenhall, Broseley, Salop.
*Williams, M., Bishton Hall, Shifnal, Salop.
*Willoughby De Broke, Lord, Compton Verney, Warwick.
*Wilson, H., Pendeford, Wolverhampton.
 Wilson, Walter, Wildmoor, Claverley, Bridgnorth, Salop.
*Woodburne, Myles, Kenwick, Ellesmere, Salop.
 Worsley-Worswick R., Normanton Hall, Hinckley.

 Yeomans, J. H., Stretton Court, Hereford.

*Zetland, Earl of, Aske, Richmond, Yorks.

www.ingramcontent.com/pod-product-compliance
Lightning Source LLC
Chambersburg PA
CBHW082012230526
45468CB00022B/2013